Cutting Power Sector Carbon Pollution:
State Policies and Programs

This document provides updated information that reflects existing state policies and programs as of August 2016. It builds on information presented in the Appendix to the State Plan Considerations Technical Support Document developed for U.S. Environmental Protection Agency Clean Power Plan proposal in June 2, 2014 (Docket ID No. EPA-HQ-OAR-2013-0602). This update does not modify that record in any way but provides, for informational purposes only, updated details about existing state policies and programs that reduce CO_2 emissions from the power sector.

Contents

List of Figures

List of Tables

List of Acronyms

ACEEE – American Council for an Energy Efficient Economy
ACP – Alternative Compliance Payment
BSER – Best System of Emissions Reduction
CACJA – Clean Air Clean Jobs Act
CCR – Cost Containment Reserve
CHP – Combined Heat and Power
CEMS – Continuous Emissions Monitoring System
CO_2 – Carbon Dioxide
CO_2e – Carbon Dioxide Equivalent
CDPHE – Colorado Department of Public Health and Environment
DOE – Department of Energy
DSIRE – Database of State Incentives for Renewable Energy
EERS – Energy Efficiency Resource Standard
EGU – Electric Generating Unit
EIA – Energy Information Administration
EM&V – Evaluation, Measurement, and Verification
EPA – Environmental Protection Agency
ERP – Electricity Resource Plan
ESCO – Energy Service Company
GDP – Gross Domestic Product
GHG – Greenhouse Gas
GW – Gigawatt (1 GW = 1,000 MW)
GWh – Gigawatt-hour (1 GWh = 1,000 MWh)
IECC – International Energy Conservation Code
IOU – Investor-Owned Utility
IRP – Integrated Resource Planning
kWh – Kilowatt-hour
LBNL – Lawrence Berkeley National Laboratory
LDC – Local Distribution Company
MERP – Metropolitan Reduction Proposal
$MMTCO_2e$ – Million Metric Tons of Carbon Dioxide Equivalent
MMBTU – Million British Thermal Units
MW – Megawatt
MWh – Megawatt-hour (1 MWh = 1,000 kWh)
NO_x – Nitrogen Oxides
PBF – Public Benefit Funds
PBI – Performance-based Incentives
RGGI – Regional Greenhouse Gas Initiative
REC – Renewable Energy Certificate
RES – Renewable Energy Standard 5

RPS – Renewable Portfolio Standard
PUC – Public Utility Commission
SO_2 – Sulfur Dioxide
VEIC – Vermont Energy Investment Corporation
WAP – Weatherization Assistance Program

I. Overview of State Climate and Energy Policies and Programs That Reduce Power Sector CO_2 Emissions

Across the nation, many states and regions have shown strong leadership in creating and implementing policies, programs, and measures that reduce CO_2 emissions from the power sector, while achieving other economic, environmental, and energy benefits. These policies and programs can serve as a strong foundation for states developing strategies to reduce greenhouse gas (GHG) emissions and for those that voluntarily choose to continue exploring options to address requirements for affected electric generating units (EGUs) under the final "Carbon Pollution Emission Guidelines for Existing Stationary Sources: Electric Utility Generating Units," also known as the Clean Power Plan.[1]

This document provides an overview of existing state activities that reduce CO_2 emissions from the power sector. Policies and programs range from market-based programs and CO_2 emissions performance standards that require CO_2 emissions reductions from EGUs, to others, such as renewable portfolio standards (RPS) and energy efficiency resource standards (EERS), that reduce CO_2 emissions by altering the mix of energy supply and reducing energy demand. States have developed their policies and programs with stakeholder input and tailored them to their own circumstances and priorities.

States vary in their regulatory structures, electricity generation, and usage patterns, while geography affects factors such as the availability of fuels, transmission networks, and seasonal energy demand. States have tailored their climate and energy policies and programs accordingly. For example, in some states, utilities are vertically integrated, meaning that the one company is responsible for electricity generation, transmission, and distribution over a given service territory. State public utility regulators have authority over these utilities. In other states, where the electric power industry has been restructured, ownership of electric generation assets has been decoupled from transmission and distribution assets, and retail customers have their choice of electricity suppliers. In states where restructuring is active (see Figure 1), state public utility regulators do not have authority to regulate the companies responsible for electricity generation, but they can regulate the electricity distribution utilities. States rely upon and have access to different fuel types and have a variety of EGU types within state borders. States are part of regional electricity grids that usually do not align with state

---

[1] On February 9, 2016, the Supreme Court stayed the Clean Power Plan. EPA is not implementing or enforcing the requirements of the rule accordingly at this time. EPA is providing technical assistance for states who choose to move forward on a voluntary basis to address the requirements of the Clean Power Plan. Available at https://www.gpo.gov/fdsys/pkg/FR-2015-10-23/pdf/2015-22842.pdf.

borders. Electricity is imported and exported by utilities across states throughout each regional grid.

Figure 1: Status of Electricity Restructuring by State

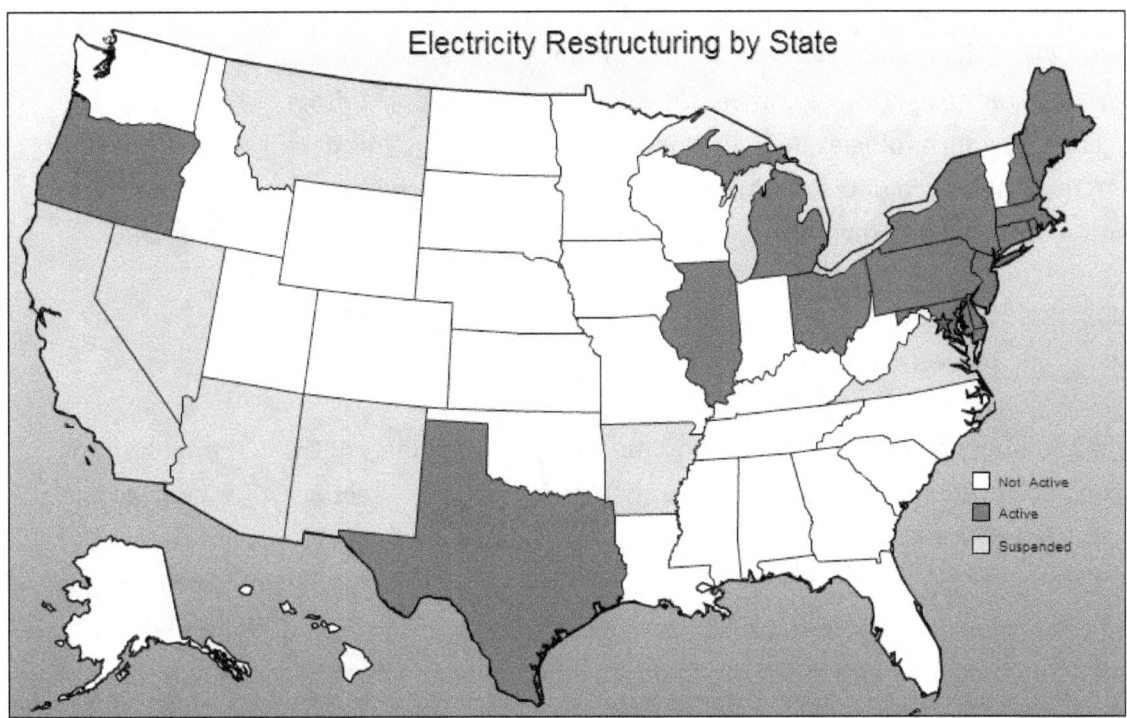

Source: "Status of Restructuring by State as of September 2010" (U.S. Energy Information Administration), accessed March 10, 2016. Available at:
http://www.eia.gov/electricity/policies/restructuring/restructure_elect.html.

States also have different economic considerations, drivers, and approaches when implementing climate change, energy efficiency, and renewable energy policies, programs, and measures. State actions may be motivated by state environmental, energy and/or economic concerns. For example, as of March 2016, 12 states and Washington, D.C., have passed legislation establishing GHG reduction goals and are using a combination of emissions limits, performance standards, energy efficiency, and renewable energy measures to achieve these goals.[2] Other state measures are motivated by public utility commission (PUC) requirements to achieve all cost-effective end-use energy efficiency improvements or by renewable energy generation requirements. Policies, programs, and measures vary from state to state in their

---

[2] States include California, Connecticut, Hawaii, Maine, Maryland, Massachusetts, Minnesota, New Jersey, Oregon, Rhode Island, Vermont and Washington. Targets are typically defined on a 1990 base year, aiming to achieve reductions of between 0 and 10 percent by 2020, although Maryland and Minnesota have chosen targets of 25 percent below 2006 levels by 2020, and 15 percent below 2005 levels by 2015 respectively. "Greenhouse Gas Emissions Targets," Center for Climate and Energy Solutions, accessed March 10, 2016. Available at: http://www.c2es.org/us-states-regions/policy-maps/emissions-targets.

implementation levels and administration. Some are administered by state agencies and others by utilities, with varying mechanisms for ensuring compliance with applicable requirements.

This document is not exhaustive and is only intended to provide background information about strategies states have used to achieve CO_2 emissions reductions in the power sector, advance end-use energy efficiency, and increase the use of renewable energy resources. For example, states may consider measures that other states have used to support other low- or zero-emitting generating technologies beyond what is addressed here. This document in no way purports to indicate or evaluate whether the state policies and programs described meet the requirements of the Clean Power Plan or a CAA section 111(d) state plan more generally.

II.   Existing State and Utility Policies, Programs, and Measures That Affect EGU CO_2 Emissions

Some state and utility policies, programs, and measures directly target EGU CO_2 emissions by creating specific limits or standards for CO_2 emissions in the power sector. Other policies and programs, such as those that advance deployment of end-use energy efficiency or renewable energy, are designed to reduce energy demand or promote an increase of supply from low- or non-GHG–emitting generating sources, which reduces CO_2 emissions from fossil fuel–fired EGUs. Many states that are aggressively pursuing climate change mitigation look to end-use energy efficiency and renewable energy first, recognizing the potential for low-cost GHG emissions reductions and the economic, reliability, and fuel diversity benefits these resources provide.

For example, according to California, "the integrated nature of the grid means that policies which displace the need for fossil generation can often cut emissions from covered sources more deeply, and more cost-effectively than can engineering changes at the plants alone, although these source-level control efforts are a vital starting point."[3] California calls its energy efficiency standards "the bedrock upon which climate policies are built" and uses renewable energy to fill any remaining energy needs.[4] On October 7, 2015, California Governor Jerry Brown signed The Clean Energy and Pollution Reduction Act of 2015, requiring California to generate half of its electricity from renewable sources by 2030 and double energy efficiency in homes, offices and factories.[5] The policies will assist California in meeting its statewide goal of reducing GHG emissions to 1990 levels by 2020, 40 percent below 1990 levels by 2030 and 80

---

[3] Mary Nichols (Chairman of California Air Resources Board), letter to EPA Administrator Gina McCarthy, December 27, 2013.

[4] Ibid.

[5] Gov. Brown signs climate change bill to spur renewable energy, efficiency standards. October 7, 2015. *LA Times*. Available at: http://www.latimes.com/politics/la-pol-sac-jerry-brown-climate-change-renewable-energy-20151007-story.html.

percent below 1990 levels by 2050.[6] As another example, Connecticut has a law that requires the state to reduce GHG emissions to 10 percent below 1990 emissions levels by 2020 and 80 percent from 2001 levels by 2050.[7] Connecticut considers energy efficiency investments, expanded renewable energy generation, and participation in the Regional Greenhouse Gas Initiative (RGGI) among its top ten strategies to reduce GHG emissions when considering cost-effectiveness and GHG emissions reduction potential.[8]

Beyond these specific policies and programs, some states implement utility planning requirements that can affect emissions both directly and indirectly. This section describes a range of existing state actions that fall into all of these categories.

A. Actions That Directly Reduce EGU CO_2 Emissions

Existing state actions that directly reduce EGU CO_2 emissions tend to fall in one of two categories: market-based emissions limits or emissions performance standards.

---

[6] Office of Governor Edmund G. Brown Jr. *Governor Brown Establishes Most Ambitious Greenhouse Gas Reduction Target in North America*, April 29, 2015. Available at: http://gov.ca.gov/news.php?id=18938.

[7] State of Connecticut, *Connecticut House Bill No. 5600: An Act Concerning Connecticut Global Warming Solutions*. Available at: http://www.cga.ct.gov/2008/ACT/PA/2008PA-00098-R00HB-05600-PA.htm.

[8] States' Section 111(d) Implementation Group Input to EPA on Carbon Pollution Standards for Existing Power Plants, Joint comments from 15 states on Carbon Pollution Standards for Existing Power Plants sent to USEPA Administrator McCarthy on December 16, 2013. Signatories include: Mary D. Nichols, Chairman of California Air Resources Board, Robert B. Weisenmiller, California Energy Commission, Michael R. Peevey, Chair of California Public Utilities Commission, Larry Wolk, MD, MSPH, Executive Director and Chief Medical Offices of Colorado Department of Public Health and Environment, Dan Esty, Commissioner of Connecticut Department of Environmental Protection, Collin O'Mara, Secretary of Delaware Department of Natural Resources and Environmental Control, Dallas Winslow, Chairman of Delaware Public Service Commission, Douglas Scott, Chair of Illinois Commerce Commission, David Littell, Commissioner of Maine Public Utilities Commission, Robert M. Summers, Secretary of Maryland Department of the Environment, Kelly Speakes-Backman, Commissioner of Maryland Public Service Commission, Ken Kimmell, Commissioner of Massachusetts Department of Environmental Protection, Mark Sylvia, Commissioner of Massachusetts Department of Energy resources, John Linc Stine, Commissioner of Minnesota Pollution Control Agency, Mike Rothman, Commissioner of Minnesota Department of Commerce, Thomas S. Burack, Commissioner of New Hampshire Department of Environmental Service, Joseph Martens, Commissioner of New York State Department of Environmental Conservation, Audrey Zibelman, Chief of New York State Public Commission, Dick Pederson, Director Oregon department of Environmental Quality, Janet Coit, Director of Rhode Island Department of Environmental Management, Marion Gold, Commissioner of Rhode Island Office of Energy resources, Deborah Markowitz, Secretary of Vermont Agency of Natural Resources, James Volz, Chairman of Vermont Public Service Board, Maia Bellon, Director of Washington State Department of Ecology. Letter hereafter referred to as "State environmental agency leaders from CA, CO, DE, IL, ME, MD, MA, MN, NH, NY, OR, RI, VT, WA, Open Letter to the EPA Administrator Gina McCarthy on Emission Standards Under Clean Air Act Section 111(d), December 16, 2013."

i. Market-based Emissions Limits

Description

An emissions budget trading program is a market-based tool for reducing pollution. The basic approach, which involves the allocation and trade of a limited number of environmental permits, has been used across environmental media, including air pollution control, clean water regulation, and land-use applications.

As of March 2016, ten states have implemented emissions budget trading programs addressing CO_2 and other GHG emissions. As shown in Figure 2 below, these include California's emissions budget trading program and the nine northeast and mid-Atlantic states participating in the Regional Greenhouse Gas Initiative (RGGI), consisting of Connecticut, Delaware, Maine, Maryland, Massachusetts, New Hampshire, New York, Rhode Island, and Vermont.[9,10]

Figure 2: States with Active Greenhouse Gas Emissions Budget Trading Programs

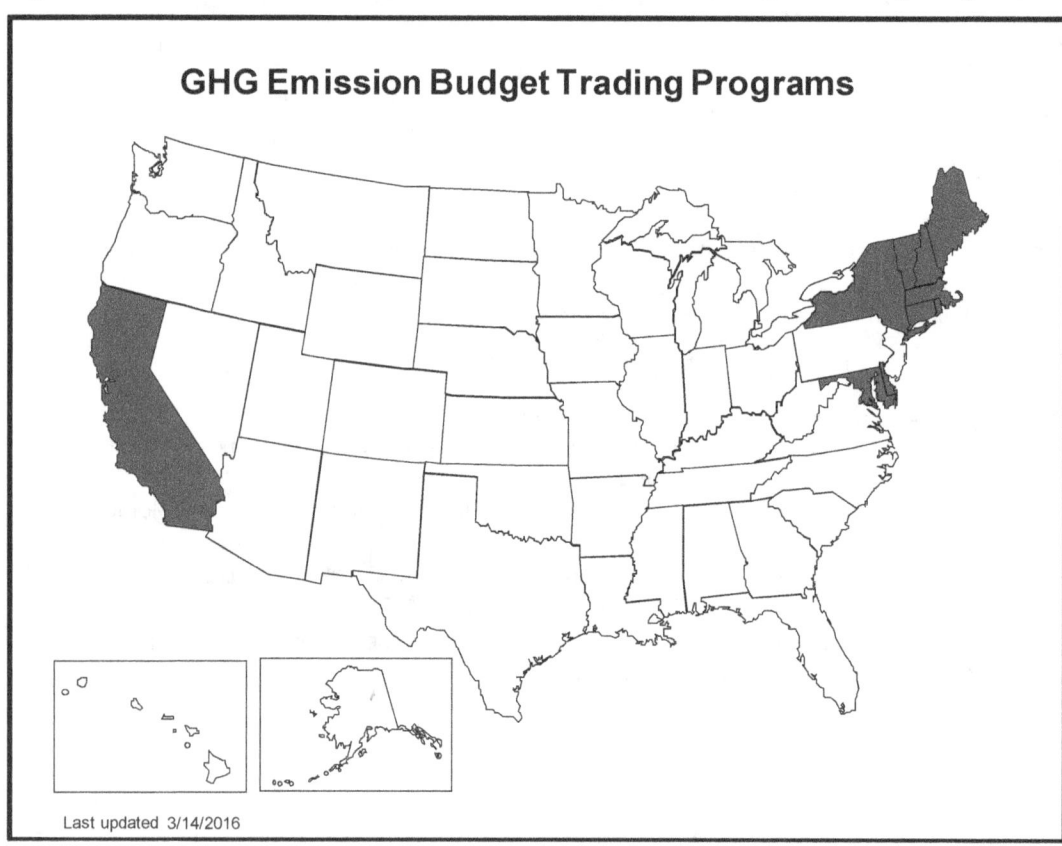

GHG Emission Budget Trading Programs

Last updated 3/14/2016

---

[9] Regional Greenhouse Gas Initiative Inc. Website Homepage, accessed March 10, 2016. Available at: http://www.rggi.org/.
[10] "Cap-and-Trade Program," California Air Resources Board, accessed March 10, 2016. Available at: http://www.arb.ca.gov/cc/capandtrade/capandtrade.htm.

Design

An emissions budget trading program establishes an aggregate limit on pollution through an emissions cap that specifies the total allowable emissions over a specified time period for all of the emissions sources subject to the program. To comply with the emissions limitation, each emissions source must surrender emissions allowances equal to its reported emissions at the end of each compliance period.

Allowances may be traded among both regulated and non-regulated parties, creating a market for emissions allowances. In turn, the allowance market establishes a price signal for emissions (a market price for emitting a unit of pollution), which triggers broad economic incentives for reducing emissions across the covered sector(s) and encourages innovation in developing emissions control strategies and new pollution control technologies.

There are several key design elements that may vary from program to program:

- Scope of coverage (e.g., sectors and types of facilities covered)

- Applicability (criteria for inclusion of emitting facilities and units in the program)

- Initial emissions budget (i.e., the aggregate emissions limitation for covered emissions sources) and emissions reduction schedule

- Flexibility provisions, in addition to ability to trade emissions allowances, including:
    - Multi-year compliance periods
    - Allowance banking
    - Offsets (e.g., project-based emissions reductions occurring outside the capped sector/sources)

- Additional provisions to mitigate price volatility and overall costs
    - Auction reserve price
    - Cost containment reserve of allowances provided for sale at set price thresholds; Once the allowance price hits a threshold, an extra supply of allowances are made available

Table 1 summarizes some of the key design elements of the RGGI and California programs.

Table 1: Comparison of RGGI and California Emissions Budget Trading Programs

Element	RGGI	California
Applicability	• All fossil fuel–fired EGUs with a capacity of 25 MW or greater.[11]	• All facilities in covered sectors, either directly emitting or distributing fossil fuels with potential combustion emissions, of at least 25,000 metric tons CO_2-equivalent (CO_2e) or greater (with no minimum[12] for emissions from imported electricity).[13]
Scope	• Facilities in electric power sector.[14]	• Facilities in electric power, large industrial sectors, and distributors of gasoline, certain diesel fuels, liquefied petroleum gas, and natural gas.[15,16]
Emissions budget	• Recently reduced 45 percent to 91 million tons of CO_2 in 2014. Beginning in 2015, the budget will decline 2.5 percent per year to 2020.[17]	• Set at 2 percent below expected 2012 emissions in 2013 (162.8 million tons of CO_2), declining by 2 percent in 2014 and 3 percent annually from 2015 (394.5 million tons of CO_2)[18] to 2020 (334.2 million tons of CO_2). [15,19]
Compliance period	• EGUs must demonstrate compliance every three years and hold allowances equal to 50 percent of reported CO_2 emissions at the end of the first two years of every three-year compliance period.[20]	• Facilities must demonstrate compliance every three years. On an annual basis, facilities must also hold allowances and offsets covering 30 percent of the previous year's emissions.[21]

[11] Regional Greenhouse Gas Initiative Inc., *Overview of RGGI CO_2 Budget Trading Program* (RGGI Inc., October, 2007). Available at: http://www.rggi.org/docs/program_summary_10_07.pdf.

[12] As of January 1, 2015, all electricity imports, regardless of the size of the generating station of origin, are covered under the emissions trading system.

[13] California Air Resources Board, *Cap and Trade Regulation Instructional Guidance, Chapter 2: Is My Company Subject to the Cap-and-Trade Regulation* (CARB, September, 2012). Available at: http://www.arb.ca.gov/cc/capandtrade/guidance/chapter2.pdf.

[14] "Regulated Sources," Regional Greenhouse Gas Initiative Inc., accessed March 10, 2016. Available at: http://www.rggi.org/design/overview/regulated_sources.

[15] California Air Resources Board, *Overview of ARB Emissions Trading Program* (CARB, October, 2011). Available at: http://www.arb.ca.gov/newsrel/2011/cap_trade_overview.pdf.

[16] California Air Resources Board, *Information for Entities That Take Delivery of Fuel for Fuels Phased into the Cap-and-Trade Program Beginning on January 1, 2015* (CARB, 2015). Available at: http://www.arb.ca.gov/cc/capandtrade/guidance/faq_fuel_purchasers.pdf.

[17] Regional Greenhouse Gas Initiative Inc., "RGGI States Make Major Cuts to Greenhouse Gas Emissions from Power Plants," Regional Greenhouse Gas Initiative Press Release (January 13, 2014). Available at: http://www.rggi.org/docs/PressReleases/PR011314_AuctionNotice23.pdf.

[18] The large cap increase in 2015 is due to the inclusion of transportation, natural gas, and other fossil fuel distributors in the emissions trading program.

[19] Center for Climate and Energy Solutions (C2ES), *California Cap and Trade* (C2ES, 2015) accessed March 10, 2016. Available at: http://www.c2es.org/us-states-regions/key-legislation/california-cap-trade.

[20] "Compliance" Regional Greenhouse Gas Initiative Inc., accessed March 10, 2016. Available at: http://www.rggi.org/market/tracking/compliance.

[21] "Regulated Sources," Regional Greenhouse Gas Initiative Inc., accessed March 10, 2016. Available at: http://www.rggi.org/design/overview/regulated_sources.

Element	RGGI	California
Allowance allocation method	• Each state distributes allowances from its established budget in an amount and manner determined by its applicable statutes and regulations. Approximately 90 percent of CO_2 allowances are distributed through auction.[22]	• Allowances are both allocated and auctioned off according to provisions established by the program. More information is available from CARB (see footnote).[15]
Cost containment provisions	• A Cost Containment Reserve (CCR) of CO_2 allowances provides a fixed additional supply of allowances that are only available if the auction price exceeds a set threshold ($4 in 2014 rising to $10 in 2017 and 2.5 percent per year to 2020).[23] • An additional five million allowances became available March 2014 when market price exceeded the current price trigger of $4 per ton.[24] • CCR allowances increase from five million in 2014 to 10 million in 2015 and beyond.[25]	• A strategic reserve is included, providing an Allowance Price Containment Reserve of 1 percent of allowances for 2013-2014, 4 percent of allowances for 2015-2017, and 7 percent of allowances for 2018-2020. Shares of allowances held in the reserve will be released at three price trigger points; $40, $45, and $50 per ton and rise by 5 percent per year including inflation.[26]
Banking	• Allows unlimited allowance banking.[27]	• Allows unlimited allowance banking, but regulated entities are subject to holding limits, which are a function of the entity's annual allowance budget.[28,29]

---

[22] Regional Greenhouse Gas Initiative Inc., "2015 Allowance Allocation." Available at: www.rggi.org/design/overview/allowance-allocation.

[23] Regional Greenhouse Gas Initiative Inc., "The RGGI CO_2 Cap," accessed March 10, 2016. Available at: http://www.rggi.org/design/overview/cap.

[24] Regional Greenhouse Gas Initiative Inc., "CO_2 Allowances Sold at $4.00 at 23rd RGGI Auction," Regional Greenhouse Gas Initiative Press Release (March 7, 2014). Available at: http://www.rggi.org/docs/Auctions/23/PR030714_Auction23.pdf.

[25] Regional Greenhouse Gas Initiative Inc., *Summary of RGGI Model Rule Changes* (Regional Greenhouse Gas Initiative, Inc., 2013). Available at: http://www.rggi.org/docs/ProgramReview/_FinalProgramReviewMaterials/Model_Rule_Summary.pdf.

[26] "California Cap on Greenhouse Gas Emissions and Market-Based Compliance Mechanisms to Allow for the Use of Compliance Instruments Issued by Linked Jurisdictions," California Code of Regulations, Title 17, §95800-96023, July 2013. Available at: http://www.arb.ca.gov/cc/capandtrade/ctlinkqc.pdf.

[27] "Cap-and-Trade Program," California Air Resources Board, accessed March 10, 2016. Available at: http://www.arb.ca.gov/cc/capandtrade/capandtrade.htm.

[28] CARB *Proposed Regulation to Implement the California Cap-and-Trade Program* (California Air Resources Board, 2010). Available at: http://www.arb.ca.gov/regact/2010/capandtrade10/capisor.pdf.

[29] The large cap increase in 2015 is due to the inclusion of transportation, natural gas, and other fossil fuel distributors in the emissions trading program.

Element	RGGI	California
Offsets	• EGUs subject to RGGI are allowed to use offsets within the RGGI region to meet 3.3 percent of their compliance obligation, increasing to 5 and 10 percent if allowance prices exceed price thresholds of $7 and $10 per allowance, respectively.[30, 31,32]	• Facilities may use domestic offsets for up to 8 percent of their compliance obligation.[33] A framework has been established to include international offsets but these are currently not allowed in the program. [34]

Authority

State and regional GHG emissions budget trading programs are authorized through individual state legislation and implemented through state regulations. For example, California implemented its emissions budget trading program under the authority of its 2006 Global Warming Solutions Act, which requires the state to reduce its 2020 GHG emissions to 1990 levels.[35] Each RGGI state has separate authorizing legislation, and in some cases, its legislation specifically directs the use of auction proceeds. For example, Maine authorized its participation in RGGI through Statute 580-A, Title 38 Chapter 3B: Regional Greenhouse Gas Initiative. This statute also requires that 100 percent of auction proceeds go toward carbon reduction and energy conservation efforts.[36] RGGI is implemented through individual state CO_2 budget trading program regulations.[37]

---

[30] Regional Greenhouse Gas Initiative Inc., "CO₂ Offsets," accessed March 10, 2016. Available at: http://www.rggi.org/market/offsets.

[31] Eligible offsets under RGGI include: landfill methane capture and destruction, sulfur hexafluoride (SF₆) reduction from power transmission, U.S. forest projects (reforestation, improved forest management, and avoided conversion) or afforestation (in Connecticut and New York only), end use energy efficiency, and agricultural manure management. "Offset Categories" Regional Greenhouse Gas Initiative, Inc., accessed March 10, 2016. Available at: http://www.rggi.org/market/offsets/categories.

[32] Regional Greenhouse Gas Initiative Inc., *Fact Sheet: RGGI Offsets*. Available at: http://www.rggi.org/docs/Documents/RGGI_Offsets_FactSheet.pdf.

[33] California Air Resources Board, *Overview of ARB Emissions Trading Program* (CARB, October, 2011). Available at: http://www.arb.ca.gov/newsrel/2011/cap_trade_overview.pdf. Offsets are initially limited to forestry, urban forestry, livestock methane capture and destruction, and destruction of ozone depleting substances. However, rice cultivation and coal mine methane are proposed for inclusion in the program. See: CARB – Potential New Compliance Offset Projects at: http://www.arb.ca.gov/cc/capandtrade/offsets/offsets.htm for more information; accessed March 10, 2016.

[34] California Air Resources Board, *Overview of ARB Emissions Trading Program* (CARB, 2011). Available at: http://www.arb.ca.gov/newsrel/2011/cap_trade_overview.pdf.

[35] Assembly Bill 32, California Global Warming Solutions Act of 2006, Division 25.5 (September 27, 2006). Available at: http://www.leginfo.ca.gov/pub/05-06/bill/asm/ab_0001-0050/ab_32_bill_20060927_chaptered.pdf.

[36] Maine revised statutes, Title 38, Chapter 3-B, section 580-B, the Regional Greenhouse Gas Initiative Act of 2007, accessed March 10, 2016. Available at: http://www.mainelegislature.org/legis/statutes/38/title38sec580-B.html.

[37] Regional Greenhouse Gas Initiative Inc., "State Statutes and Regulations," accessed March 10, 2016. Available at: http://www.rggi.org/design/regulations.

The state regulatory authority issues individual authorizations to emit a specific quantity of emissions ("allowances"), which represent one (metric or short) ton of a pollutant, in an amount no greater than the established emissions budget.

Obligated Parties

Obligated parties in emissions budget trading programs are generally the covered emissions sources. The emissions sources are responsible for surrendering emissions allowances equal to their reported emissions at the end of each compliance period. For example, as stated above, RGGI covers fossil fuel–fired EGUs 25 megawatts or larger in size.[38] The California emissions budget trading program covers electricity generators, distributors of transportation, natural gas, and other fuels, and industrial facilities with emissions[39] greater than 25,000 metric tons CO_2-e. The program also covers all importers of electricity.[40]

Measurement and Verification

Emissions budget trading programs include requirements for emissions monitoring and reporting by affected emissions sources, holding and transfer of allowances, and surrender of allowances (and offset allowances or credits) in an amount equal to reported emissions. Allowance surrender in an amount equal to reported emissions is often referred to, generally, as the program "compliance obligation."

For example, EGUs subject to the RGGI program must report CO_2 emissions quarterly pursuant to state regulations, which are generally consistent with EPA regulations for reporting of CO_2 emissions from EGUs under 40 CFR 75.[41] Emissions are reported quarterly to EPA, using the Emissions Collection and Monitoring Plan System (ECMPS), and data is transferred to the RGGI CO_2 Allowance Tracking System (RGGI COATS). GHG emissions reporting for affected sources under the California program is addressed through the California mandatory GHG reporting regulations, using a modified version of the reporting platform administered through the EPA Greenhouse Gas Reporting Program.[42] Affected emissions sources must report emissions annually and provide third party verification of reported emissions.

---

[38] "Regulated Sources," Regional Greenhouse Gas Initiative Inc., accessed March 10, 2016. Available at: http://www.rggi.org/design/overview/regulated_sources.

[39] Fossil fuel distributors are liable for combustion emissions that occur downstream of their operations.

[40] California Air Resources Board, *Overview of ARB Emissions Trading Program* (CARB, 2011). Available at: http://www.arb.ca.gov/newsrel/2011/cap_trade_overview.pdf.

[41] Regional Greenhouse Gas Initiative Inc., *Overview of RGGI CO_2 Budget Trading Program* (RGGI Inc., 2007). Available at: http://www.rggi.org/docs/program_summary_10_07.pdf.

[42] California Air Resources Board, *Overview of ARB Emissions Trading Program* (CARB, 2011). Available at: http://www.arb.ca.gov/newsrel/2011/cap_trade_overview.pdf.

Penalties for Non-compliance

Failure to submit allowances in an amount equal to reported emissions result in automatic emissions penalties in the form of additional allowance submission requirements (e.g., three-to-one submission requirements to account for any shortfall in RGGI, and a four-to-one submission requirement for any shortfall under the California program). States may also apply other administrative fines and penalties, pursuant to their implementing regulations.

Implementation Status

The RGGI program was established in 2009. From 2009 through 2014, the nine current RGGI participating states invested auction proceeds of more than $1.3 billion in programs that lower costs for energy consumers and reduce CO_2 emissions, including more than $750 million in energy efficiency programs and more than $300 million in renewable energy.[43] The participating RGGI states estimate that all of their investments are providing benefits of $4.67 billion in lifetime energy savings to energy consumers in the region.[44]

Between 2005, when agreement to implement RGGI was first announced, and 2014, power sector CO_2 emissions in the RGGI participating states fell by more than 40 percent while GDP in the region grew by more than 8 percent (see Figure 3).[45]

---

[43] Regional Greenhouse Gas Initiative Inc., *Investment of RGGI Proceeds Through 2014* (RGGI Inc., 2016). Available at: https://www.rggi.org/docs/ProceedsReport/RGGI_Proceeds_Report_2014.pdf. Programs include residential, commercial, and industrial programs. Of the $1.37 billion in auction proceeds invested by RGGI participating states through 2014, approximately 58 percent supported end-use energy efficiency programs and approximately 23 percent supported renewable energy programs.

[44] Fossil fuel distributors are liable for combustion emissions that occur downstream of their operations.

[45] Regional Greenhouse Gas Initiative Inc., *Investment of RGGI Proceeds Through 2014* (RGGI Inc., 2016). Available at: https://www.rggi.org/docs/ProceedsReport/RGGI_Proceeds_Report_2014.pdf.

By contrast, total U.S. power sector CO_2 emissions fell by 15 percent during the same period of time. See 2015 U.S. Greenhouse Gas Inventory for more detail: U.S. EPA, *Inventory of U.S. Greenhouse Gas Emissions and Sinks: 1990-2013* (U.S. Environmental Protection Agency, 2015), Available at: https://www3.epa.gov/climatechange/ghgemissions/usinventoryreport/archive.html.

Figure 3: Historical GDP and Greenhouse Gas Emissions in the RGGI Region

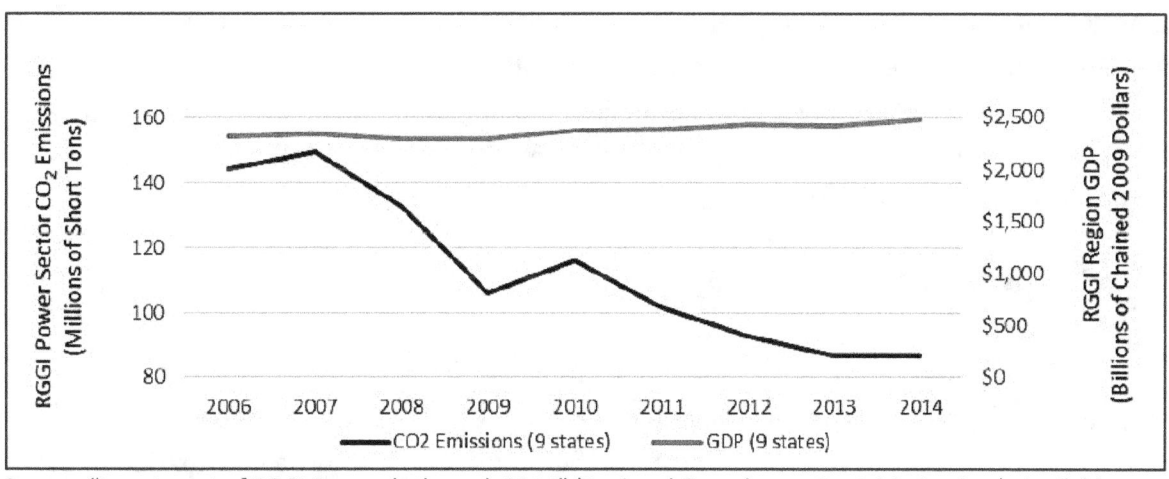

Source: "Investment of RGGI Proceeds through 2014" (Regional Greenhouse Gas Initiative, Inc.). Available at: https://www.rggi.org/docs/ProceedsReport/RGGI_Proceeds_Report_2014.pdf.

The RGGI program, which began in 2009, was not a primary driver for these emissions reductions in RGGI states, but the lower emissions led participating states to adjust the multi-state CO_2 emissions limit.[46] In January 2014, the RGGI participating states lowered the overall allowable CO_2 emissions level in 2014 by 45 percent, setting a multi-state CO_2 emissions limit for affected EGUs of 91 million short tons of CO_2 in 2014, falling to 78 million short tons of CO_2 by 2020, approximately 50 percent below 2005 levels.[47,48] Actual 2014 emissions were 85 million short tons of CO_2, slightly below the cap.[49]

The California economy-wide market-based GHG emissions budget trading program, which addresses GHG emissions from multiple sectors, was implemented in 2012 with emissions limits beginning in 2013.[50,51] While California's emissions budget trading program, like its state

[46] U.S. Energy Information Administration (EIA), *Lower emissions cap for Regional Greenhouse Gas Initiative takes effect in 2014* (EIA, 2014). Available at: http://www.eia.gov/todayinenergy/detail.cfm?id=14851. The first three-year control period under RGGI, establishing CO_2 emissions limits for EGUs, began on January 1, 2009. Low gas prices, increased renewables, decreased electric demand and weather are considered four primary drivers of the reductions through 2010, as reported by Environment Northeast in May 2011.

[47] Regional Greenhouse Gas Initiative Inc., "RGGI States Make Major Cuts to Greenhouse Gas Emissions from Power Plants," Regional Greenhouse Gas Initiative Press Release (January 13, 2014). Available at: http://www.rggi.org/docs/PressReleases/PR011314_AuctionNotice23.pdf.

[48] Regional Greenhouse Gas Initiative Inc., "The RGGI CO_2 Cap," accessed March 10, 2016. Available at: http://www.rggi.org/design/overview/cap.

[49] Regional Greenhouse Gas Initiative Inc., *Annual Report on the Market for RGGI CO_2 Allowances: 2014*, (Regional Greenhouse Gas Initiative, May 2015). Available at: http://rggi.org/docs/Market/MM_2014_Annual_Report.pdf. Cumulative CO_2 emissions for the second control period (2012-2014) rose from 179 to 264 million short tons throughout 2014, a difference of 85 million short tons.

[50] "Cap-and-Trade Program," California Air Resources Board, accessed March 10, 2016. Available at: http://www.arb.ca.gov/cc/capandtrade/capandtrade.htm.

[51] The California program was developed in coordination with U.S. state and Canadian province WCI partners.

emissions limit, is multi-sector in scope, the state projects that the emissions trading program and related complementary measures will reduce power sector GHG emissions to less than 80 million metric tons of CO_2-e by 2025, a 25 percent reduction from 2005 power sector emissions levels.[52] Prior to the implementation of the emissions trading program, California reports that it reduced power sector CO_2 emissions by 16 percent from 2005 to a 2011-2013 averaging period, a reduction of 16 million metric tons of CO_2-e.[53]

ii. CO_2 Emissions Performance Standards

Description

CO_2 emissions performance standards can apply either directly to EGUs or to the local distribution company (LDC) that sells electricity to the customers. (For more information about how electricity is generated and distributed, see Chapter 2 of the Regulatory Impact Analysis).

---

[52] State environmental agency leaders from CA, CO, DE, IL, ME, MD, MA, MN, NH, NY, OR, RI, VT, WA, Open Letter to the EPA Administrator Gina McCarthy on Emission Standards under Clean Air Act Section 111(d), December 16, 2013. Available at: http://www.eenews.net/assets/2013/12/16/document_gw_06.pdf.
Preliminary California Air Resources Board analyses, based in part on CARB 2008 to 2012 Emissions for Mandatory GHG reporting Summary (2013), cited in this letter.
[53] California Greenhouse Gas Inventory. 2000-2013 Inventory by Economic Sector – Full Detail. Available at: http://www.arb.ca.gov/cc/inventory/data/tables/ghg_inventory_sector_all_2000-13_20150831.pdf.

Figure 4: States with Greenhouse Gas Performance Standards

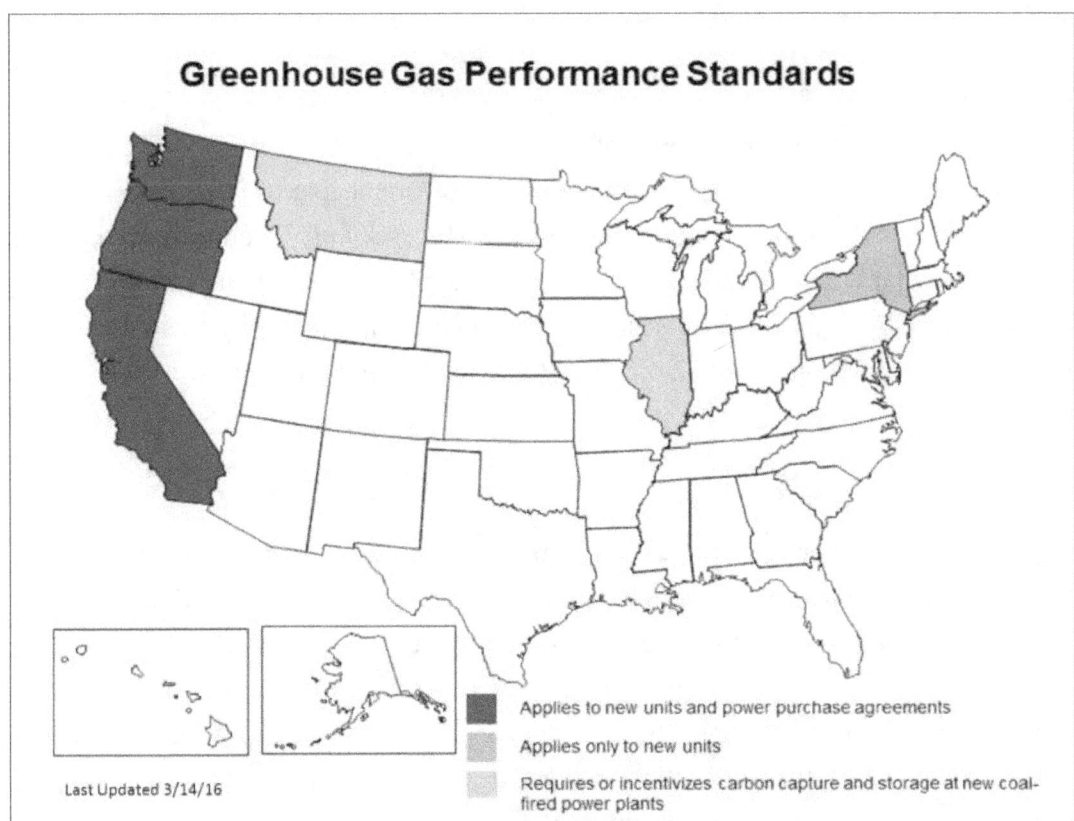

As shown above in Figure 4, , as of March 2016, four states—California, New York, Oregon, and Washington—have enacted mandatory GHG emissions standards that impose enforceable emissions limits on new and/or expanded electric generating units.[54] Three states—California, Oregon, and Washington—have enacted mandatory GHG emissions performance standards that set an emissions rate for electricity purchased by electric utilities.[54] In addition to these states, Illinois and Montana have policies to incentivize or require new coal plants to capture at least 50 percent of their CO_2 emissions.[54]

---

[54] California Energy Commission, California SB 1368, Chapter 598: Emission Performance Standards, September 29, 2006. Available at: http://www.energy.ca.gov/emission_standards/. New York Department of Environmental Conservation, Part 251: CO_2 Performance Standards for Major Electric Generating Facilities, June 12, 2012. Available at: https://govt.westlaw.com/nycrr/Browse/Home/NewYork/NewYorkCodesRulesandRegulations?guid=I5d3c9d90eaf b11e19f380000845b8d3e&originationContext=documenttoc&transitionType=Default&contextData=(sc.Default). Oregon Department of Energy, Oregon's Carbon Dioxide Emission Standards for New Energy Facilities (Oregon Department of Energy, 2010). Available at: http://www.oregon.gov/energy/Siting/docs/Reports/CO2Standard.pdf. Washington State Legislature, Chapter 80.70 RCW: Carbon Dioxide Mitigation. Available at: http://apps.leg.wa.gov/rcw/default.aspx?cite=80.70&full=true. Illinois General Assembly, Public Act 095-1027, SB1987, Clean Coal Portfolio Standard Law, January 12, 2009. Available at: http://ilga.gov/legislation/publicacts/95/PDF/095-1027.pdf. Montana State Legislature, H.B.0025.05, An Act Generally Revising the Electric Utility Industry and Customer Choice Laws, May 14, 2007. Available at: http://leg.mt.gov/bills/2007/billpdf/HB0025.pdf.

Design

States have implemented three different types of CO_2 performance standards that affect EGUs and/or LDCs differently. The first requires power plant emissions per electricity generated to be less than or equivalent to an established standard and is directly applicable to EGUs. The second type places conditions on the emissions attributes of electricity procured by electric utilities. It consists of standards that are applicable to LDCs that provide electricity to retail customers. A third type requires that new coal-fired power plants must capture and store a specific percentage of CO_2 emissions. Table 2 provides state examples for each of the types of CO_2 performance standards.

Authority

In some states, programs are regulated through the Public Utilities Commission (California, Montana).[55,56] Oregon's program is regulated through the Oregon Energy Facility Siting Council.[57] New York's program is regulated through the Department of Environmental Conservation.[58] Washington's program is regulated through two different sets of entities depending on the ownership of the utilities. The Washington Department of Community, Trade & Economic Development (CTED) is responsible for updating the emissions performance standard every five years.[59] In addition, the Washington Energy Facility Site Evaluation Council (EFSEC) is in charge of evaluating and licensing state power plants.[60] Illinois's program is regulated by the Illinois Commerce Commission.[61]

---

[55] California Energy Commission, "California SB 1368, Chapter 598: Emission Performance Standards" (September 29, 2006). Available at: http://www.energy.ca.gov/emission_standards/.

[56] Center for Climate and Energy Solutions (C2ES), "Standards and Caps for Electricity GHG Emissions" (C2ES, 2015) accessed March 10, 2016. Available at: http://www.c2es.org/us-states-regions/policy-maps/electricity-emissions-caps.

[57] Oregon Department of Energy, *Oregon's Carbon Dioxide Emission Standards for New Energy Facilities* (Oregon Department of Energy, 2010). Available at: http://www.oregon.gov/energy/Siting/docs/Reports/CO2Standard.pdf.

[58] New York Department of Environmental Conservation, "DEC Adopts Ground-Breaking Power Sector Regulations to Analyze Possible Environmental Impacts and Limit CO_2 Emissions from Power Plants," (NY DEC, 2012). Available at: http://www.dec.ny.gov/press/83269.html.

[59] Regulatory Assistance Project, "Emissions Performance Standards in Selected States" (RAP, 2009). Available at: http://www.raponline.org/docs/RAP_Simpson_EPSResearchBrief_2009_11_13.pdf.

[60] State of Washington, "Energy Facility Site Evaluation Council" (State of Washington, 2015), accessed March 10, 2016. Available at: http://www.efsec.wa.gov/default.shtm.

[61] State of Illinois, "Illinois SB 1987: Clean Coal Portfolio Standard Law" (January 12, 2009). Available at: http://www.c2es.org/docUploads/IL%20SB1987%20Coal.pdf.

Obligated Parties

The emissions performance standard can apply either directly to EGUs or to the local distribution company (LDC) that sells electricity to the customer.

Measurement and Verification

Obligated parties must measure and report on electricity generation and CO_2 emissions on a regular basis to verify their compliance with the standard. The reporting requirements and timing varies from state to state and are typically set by the agency that oversees the program as described under authority above.

Table 2 provides an overview of different CO_2 performance standards, while Table 3 provides examples regarding measurement and verification requirements across California, New York, Oregon, and Washington.

Table 2: Examples of State CO$_2$ Performance Standards

What It Does	State Examples
Requires power plant emissions per electricity generated to be less than or equivalent to the established standard; applies to EGUs	• New York (Part 251, 2012) – New or expanded baseload plants (25 MW and larger) must meet an emissions rate of either 925 lbs. CO$_2$/MWh (output based) or 120 lbs CO$_2$/MMBTU (input based). Non-baseload plants (25 MW and larger) must meet an emissions rate of either 1450 lbs. CO$_2$/MWh (output based) or 160 lbs. CO$_2$/MMBTU (input based).[62] • Oregon (HB 3283; 1997, 2007) – New natural gas-fired power plants (baseload and non-baseload) must meet an emissions rate of 675 lbs. CO$_2$/MWh. Cogeneration and offsets may be used to comply with the emissions standard.[63] Baseload power plants must meet an emissions rate of 1,100 lbs. CO$_2$/MWh.[64] • Washington (RCW 80-70-010; 2004, SB 6001) – New EGUs 25 MW and larger must have an approved CO$_2$ mitigation plan that results in mitigation of 20 percent of the total CO$_2$ emissions over the life of the facility; includes modifications to existing EGUs that result in an increase in CO$_2$ emissions of 15 percent or more. The CO$_2$ mitigation plan may include one or more of a list of eligible measures (includes indirect measures, such as EE/RE and offsets).[65] Baseload power plants must meet an emissions rate of 1,100 lbs. CO$_2$/MWh.[66]
Places conditions on the emissions attributes of electricity procured by electric utilities; applies to LDCs	• California (SB 1368; 2006) – Electric utilities may only enter into long-term power purchase agreements for baseload power if the electric generator supplying the power has a CO$_2$ emissions rate that does not exceed that of a natural gas combined cycle plant. The California Energy Commission promulgated regulations establishing an emissions rate of 1,100 lbs. CO$_2$/MWh.[67] By comparison, the average emissions rate of gas plants in the U.S. is 945 lbs. CO$_2$/MWh, while the average emissions rate of pulverized coal plants is 2,154 lbs. CO$_2$/MWh.[68] • Oregon (HB SB 101; 2009) and Washington (SB 6001; 2007) – Electric utilities may only enter into long-term power purchase agreements for baseload power if the electric generator supplying the power has a CO$_2$ emissions rate of 1,100 lbs. CO$_2$/MWh or less.

---

[62] New York Department of Environmental Conservation, "DEC Adopts Ground-Breaking Power Sector Regulations to Analyze Possible Environmental Impacts and Limit CO$_2$ Emissions from Power Plants," (NY DEC, 2012). Available at: http://www.dec.ny.gov/press/83269.html.

[63] Oregon Department of Energy, *Oregon's Carbon Dioxide Emission Standards for New Energy Facilities* (Oregon Department of Energy, 2010). Available at: http://www.oregon.gov/energy/Siting/docs/Reports/CO2Standard.pdf.

[64] Center for Climate and Energy Solutions (C2ES), Standards and Caps for Electricity GHG Emissions (C2ES, 2015), accessed March 10, 2016. Available at: http://www.c2es.org/us-states-regions/policy-maps/electricity-emissions-caps.

[65] Washington State Legislature, Chapter 80.70 RCW: Carbon Dioxide Mitigation. Available at: http://apps.leg.wa.gov/rcw/default.aspx?cite=80.70&full=true.

[66] Center for Climate and Energy Solutions (C2ES), Standards and Caps for Electricity GHG Emissions (C2ES, 2015), accessed March 10, 2016. Available at: http://www.c2es.org/us-states-regions/policy-maps/electricity-emissions-caps.

[67] California Energy Commission, California SB 1368, Chapter 598: Emission Performance Standards, September 29, 2006, accessed March 10, 2016. Available at: http://www.energy.ca.gov/emission_standards/.

[68] U.S. EPA, *eGRID 2010 data files* (U.S. Environmental Protection Agency, 2014), accessed March 10, 2016. Available at: https://www.epa.gov/energy/egrid. See "Download all eGRID files (1996-2012) (ZIP)" link.

What It Does	State Examples
Requires that new coal-fired power plants must capture and store a specific percentage of CO_2 emissions	• Illinois (SB 1987; 2009) – Illinois utilities and retailers must purchase at least 5 percent of their electricity from Clean Coal Facilities in 2015 and beyond. To be designated a Clean Coal Facility, new coal-fired power plants must capture and store 50 percent of carbon emissions from 2009-2015, 70 percent for 2016-2017, and 90 percent after 2017.[69] • Montana (HB 25; 2007) – The Public Service Commission may not approve new plants constructed after January 2007 that are primarily coal-fired unless at least 50 percent of the plant's CO_2 emissions are captured and stored. These requirements apply to formerly restructured utilities in the state. Northwest Energy is the only utility subject to this requirement, which serves about two-thirds of Montana.

Table 3: Examples of Measurement and Verification Requirements for CO_2 Performance Standards

State	Measurement and Verification Details
California	• The California PUC is responsible for approving any long-term financial commitment by an electric utility and must adopt rules to enforce these requirements as well as verification procedures.[70]
New York	• CO_2 emissions regulations require recordkeeping, monitoring, and reporting consistent with existing state and federal regulations. • Each applicable emissions source must install Continuous Emissions Monitoring Systems (CEMS) subject to federal CO_2 reporting requirements for 40 CFR part 75, successfully complete certification tests, and record, report, and quality assure the data from the CEMS. • The owner or operator must report the CO_2 mass emissions data and heat input data on a semi-annual basis to the Department of Environmental Conservation. • On a quarterly basis, the owner or operator must report all of the data and information required in either 40 CFR part 60 or subpart H of 40 CFR part 75.[71]

---

[69] Illinois General Assembly, Public Act 095-1027, SB1987, Clean Coal Portfolio Standard Law, January 12, 2009. Available at: http://ilga.gov/legislation/publicacts/95/PDF/095-1027.pdf.

[70] "SB 1368 Emission Performance Standards," California Energy Commission, accessed March 10, 2016. Available at: http://www.energy.ca.gov/emission_standards/.

[71] New York Department of Environmental Conservation, "DEC Adopts Ground-Breaking Power Sector Regulations to Analyze Possible Environmental Impacts and Limit CO_2 Emissions from Power Plants" (NY DEC, 2012). Available at: http://www.dec.ny.gov/press/83269.html.

State	Measurement and Verification Details
Washington	• Mitigation projects must be approved by the appropriate council, department, or authority, and made a condition of the proposed and final site certification agreement or order of approval. • Direct investment projects are approved if they provide reasonable certainty that the performance requirements of the projects will be achieved and that they were implemented after July 1, 2004. • For facilities under the jurisdiction of a council, the implementation of a carbon dioxide mitigation project, other than purchase of carbon credits, is monitored by an independent entity for conformance with the performance requirements of the carbon dioxide mitigation plan. The independent entity shares the project monitoring results with the council. • For facilities under jurisdiction of the department or authority, the implementation of a carbon dioxide mitigation project, other than a purchase of carbon credits, is monitored by the department or authority issuing the order of approval.[72]
Oregon	• It is up to the Council during the certificate application phase to determine the gross CO_2 emissions over a 30-year lifetime of the proposed facility to determine whether it meets the CO_2 performance standard. • During the operation phase of approved facilities, there are CO_2 reporting requirements to the Oregon Department of Environmental Quality and US EPA. • New facilities must pass a 100-hour test in their first year of operation to show they meet the performance standards.[73]

Penalties for Noncompliance

For policies that affect new electric generating units, utilities must prove any proposed units are in compliance at the time of permitting. In Oregon, if facilities do not meet the performance standard in their first year of operation during a 100-hour test,[74] they must purchase offsets to account for any excess emissions.[75]

---

[72] Washington State Legislature, Chapter 80.70 RCW: Carbon Dioxide Mitigation. Available at: http://apps.leg.wa.gov/rcw/default.aspx?cite=80.70&full=true.

[73] Oregon Department of Energy, *Oregon's Carbon Dioxide Emission Standards for New Energy Facilities* (Oregon Department of Energy, 2010). Available at: http://www.oregon.gov/energy/Siting/docs/Reports/CO2Standard.pdf.

[74] During the first year of operation new power plants test their equipment to ensure compliance with standards for commercial equipment. Initial CO_2 performance requirements can be validated during this test.

[75] Oregon Department of Energy, *Oregon's Carbon Dioxide Emission Standards for New Energy Facilities*. Available at: http://www.oregon.gov/energy/Siting/docs/Reports/CO2Standard.pdf.

Between 2007, when California enacted the performance standard and 2013, California's carbon emissions rates fell from approximately 860 lbs. CO_2e/MWh for all generation (considering both in-state and imported power) to approximately 710 lbs. CO_2e/MWh.[76]

B. Energy Efficiency Policies, Programs, and Measures

Demand-side energy efficiency policies and programs reduce utilization of EGUs and avoid GHG emissions associated with electricity generation. These electricity demand reductions can be achieved through enabling policies that incentivize investment in demand-side energy efficiency improvements by overcoming market barriers that otherwise prevent these investments. Barriers include a lack of information on energy efficient options, high transaction costs, split-incentives, lack of product availability, and perceptions of organizational risks. Reducing electricity demand also reduces the associated transmission and distribution losses that occur across the grid between the sites of electricity generation and the end use.

Demand-side energy efficiency is considered a central part of climate change mitigation in states that currently have legislated GHG targets,[77] accounting for roughly 35 percent to 70 percent of expected reductions of these states' power sector emissions.[78] For example, under California's Climate Change Scoping Plan, the state projects reductions of 21.9 million metric tons of carbon dioxide equivalent ($MMTCO_2e$) in 2020 from energy efficiency programs targeting electricity reductions. Taking into account projected reductions of 21.3 $MMTCO_2e$ from California's RPS and the expected 2.1 $MMTCO_2e$ reduction from the Million Solar Roofs program, energy efficiency makes up 48 percent of power sector reductions based on California's Climate Change Scoping Plan.[79] Another state, Washington, projects to reduce 9.7 $MMTCO_2e$ from energy efficiency measures in 2020 through a mix of new and existing programs. Taking into account expected reductions of 4.1 $MMTCO_2e$ from Washington's RPS, energy efficiency makes up 70 percent of expected emissions reductions from stationary energy within the state.[80]

---

[76] California Air Resources Board, *California Greenhouse Gas Emissions for 2000 to 2013 – Trends of Emissions and Other* Indicators (June 2015). Available at: http://www.arb.ca.gov/cc/inventory/pubs/reports/ghg_inventory_trends_00-13%20_10sep2015.pdf.

[77] States with legislated GHG targets include California, Connecticut, Hawaii, Maine, Maryland, Massachusetts, Minnesota, New Jersey, Oregon, Vermont, and Washington.

[78] These reduction target ranges are based on a review of state GHG reduction laws in states with legislated GHG targets.

[79] California Air Resources Board, *Climate Change Scoping Plan* (December 2009). Available at: http://www.arb.ca.gov/cc/scopingplan/document/adopted_scoping_plan.pdf.

[80] Washington Department of Ecology, *Growing Washington's Economy in a Carbon-Constrained World* (December 2008). Available at: https://fortress.wa.gov/ecy/publications/publications/0801025.pdf.

States have employed a variety of strategies to increase investment in demand-side energy efficiency technologies and practices, including (1) energy efficiency resource standards, (2) demand-side energy efficiency programs, (3) building energy codes, (4) appliance standards, and (5) tax credits. Each of these strategies is described below.

i. Energy Efficiency Resource Standards

Description

Energy Efficiency Resource Standards (EERS) set multiyear targets for energy savings that utilities or third-party program administrators typically meet through customer energy efficiency programs but also through other approaches, such as peak demand reductions, building codes and combined heat and power (CHP). An EERS can apply to retail distributors of either electricity or natural gas, or both, depending on the state. To date, 24 states have mandatory EE requirements in place, two states have voluntary targets, and two more states allow EE to be used to meet part of a mandatory RPS, for a total of at least 28 states with some type of EE requirement or goal.[81,82]

Policy Mechanics

Design

EERS design and implementation details vary by state, and may be expressed as a percentage reduction in annual retail electricity sales, as a percentage reduction in retail electricity sales growth, or as a specific electricity savings amount over a long-term period. A typical EERS sets multiyear targets for energy savings that drive investment in EE programs implemented by utilities or third party administrators. Over the compliance period, an EERS reduces electricity demand by a target amount that utilities must meet. As a result, an EERS indirectly affects utility CO_2 emissions by reducing the use of fossil fuel–fired EGUs.

---

[81] "State Energy Efficiency Resource Standards (EERS)" (American Council for an Energy-Efficient Economy, April 2014). Available at: http://www.aceee.org/files/pdf/policy-brief/eers-04-2014.pdf.

[82] New Hampshire has been included in this total since its mandatory EERS has been legislated, although the first year of the program is 2018. Delaware and Florida were not included in the totals. Delaware has enacted legislation to create an EERS, but final regulations have not yet been promulgated (Database of State Incentives for Renewables & Efficiency, January 2015). Available at: http://programs.dsireusa.org/system/program/detail/4510. Florida has enacted an EERS, but program funding to date is considered to be "...far below what is necessary to meet targets" ("State Energy Efficiency Resource Standards [EERS]," American Council for an Energy-Efficient Economy, April 2014). Available at: http://www.aceee.org/files/pdf/policy-brief/eers-04-2014.pdf. Ohio's EERS, while included in the total, was frozen for two years beginning in 2015. Cumulative targets will increase again from 2017 (Database of States Incentives for Renewables & Efficiency, December 2014). Available at: http://programs.dsireusa.org/system/program/detail/4542.

Authority

Most state EERS policies are established through legislation. However, there are several instances in which they have been established by PUC orders under broader statutory authority, such as by setting quantitative targets consistent with the achievement of "all cost-effective energy efficiency."[83]

Obligated Parties

Retail electricity suppliers, which are utilities that sell electricity to customers for end-use purposes, are the obligated parties under an EERS.

Measurement and Verification

PUCs generally oversee EERS. Retail electricity suppliers comply with EERS requirements by developing a portfolio of end-use energy efficiency programs that encourage electric utility customers to invest in more energy efficient technologies and practices as described below. Transmission and distribution infrastructure improvements may also count toward EERS programs in some states.[84] PUCs typically rely on independent program evaluators to perform evaluation, measurement, and verification (EM&V) activities that estimate the incremental annual and cumulative energy savings attributable to the programs.[85] These estimates are typically the basis for compliance reports submitted by retail electricity suppliers. See Table 4 for examples of penalties for program noncompliance.

---

[83] Ernest Orlando, *Benefits and Costs of Aggressive Energy Efficiency Programs and the Impacts of Alternative Sources of Funding: Case Study of Massachusetts* (Lawrence Berkeley National Laboratory, August 2010). Available at: http://emp.lbl.gov/sites/all/files/REPORT%20lbnl-3833e.pdf. An important policy driver for EE programs in six states is a statutory requirement for utilities to acquire "all cost-effective energy efficiency." This policy typically requires utilities and other program administrators to pursue energy efficiency up to the point at which it is no longer cost effective, as defined by cost-benefit tests and procedures REQUIRED by state PUCs. States with all-cost effective energy efficiency policies include: CA, CT, MA, RI, VT, WA. For MA, this goals has translated into achieving annual electric energy savings equivalent to a 2.4 percent reduction in retail sales from energy efficiency programs in 2012.

[84] For example, Ohio allows transmission and distribution infrastructure improvements to count toward its EERS (Database of State Incentives for Renewables & Efficiency, December 2014). Available at: http://programs.dsireusa.org/system/program/detail/4542.

[85] EM&V refers to set of techniques and approaches used to estimate the quantity of energy savings from an EE program or policy. Since energy savings cannot be directly measured, efficiency program impacts are estimated by taking the difference between: (a) actual energy consumption after efficiency measures are installed, and (b) the energy consumption that would have occurred during the same period had the efficiency measures not been installed (i.e., the baseline).

Penalties for Noncompliance

If the obligated parties do not demonstrate compliance with the EERS, they may face financial penalties. The existence and amount of penalties varies across the states. Table 4 provides examples of financial penalties in three states, Pennsylvania, Ohio and Illinois.

Table 4: Examples of Penalties for Noncompliance

State	Direct Financial Penalties
Pennsylvania	Failure to achieve the requisite reductions in electricity consumption and peak demand during Phase 1 results in one-time fines from $1 million to $20 million. Failure to file a plan with the public utilities commission is also punishable by a fine of $100,000 per day. Costs associated with any such fines may not be passed on to ratepayers.[86]
Ohio	Failure to comply with energy efficiency or peak demand reduction requirements results in the state public utilities commission assessing a forfeiture upon the utility, to be credited to the Advanced Energy Fund. The amount of the forfeiture is either: an amount, per day per under-compliance or non-compliance, not greater than $10,000 per violation; or an amount equal to the then existing market value of one renewable energy credit (REC)[87] per megawatt hour of under-compliance or noncompliance.[88]
Illinois	For both natural gas and electric utilities, failure to submit an energy reduction plan will result in a fine of $100,000 per day until the plan is filed. This penalty is deposited in the Energy Efficiency Trust Fund and may not be recovered by ratepayers. If an electric utility fails to comply with its plan after two years, it must make a contribution to the Low-Income Home Energy Assistance Program (LIHEAP). Large utilities (those with more than 2,000,000 customers on December 31, 2005) must contribute $665,000, and medium utilities (those with between 100,000 and 2,000,000 customers) must contribute $335,000. Utilities that fail to meet their plans again after the third year must make another contribution to the fund ($665,000 for large utilities and $335,000 for medium utilities). After three years of non-compliance, the Illinois Power Agency shall assume control over energy efficiency incentive programs. For natural gas utilities that fail to meet their efficiency plans after three years, large utilities (those with more than 1,500,000 customers on December 31, 2008) must pay $600,000 into LIHEAP, medium utilities (those with 500,000-1,500,000 customers on December 31, 2008) must pay $400,000, and small utilities (those with 100,000-500,000 customers on December 31, 2008) must pay $200,000. If a utility fails to meet the standard for two consecutive three-year planning periods, the Illinois Commerce Commission will transfer responsibility of the utility's energy efficiency programs to an independent administrator.[89]

---

[86] "Energy Efficiency and Conservation Requirements for Utilities: Pennsylvania" (Database of State Incentives for Renewables & Efficiency, June 2015). Available at: http://programs.dsireusa.org/system/program/detail/4514.

[87] RECs represent the non-energy attributes, including all the environmental attributes, of electricity generation from renewable energy sources. RECs are typically issued in single MWh increments. See the section on Renewable Portfolio Standards for more detail.

[88] "Energy Efficiency Portfolio Standard: Ohio" (Database of State Incentives for Renewables & Efficiency, December 2014). Available at: http://programs.dsireusa.org/system/program/detail/4542.

[89] "Energy Efficiency Standard: Illinois" (Database of State Incentives for Renewables & Efficiency, February 2016). Available at: http://programs.dsireusa.org/system/program/detail/4501.

As of March 2016, 24 states have an EERS program in place, while at least two have EE targets or goals that are voluntary at this time (see Figure 5). In addition, two states have renewable portfolio standard that allow the option for energy efficiency to meet requirements.[90]

Figure 5: Status of Energy Efficiency Resource Standards by State[91]

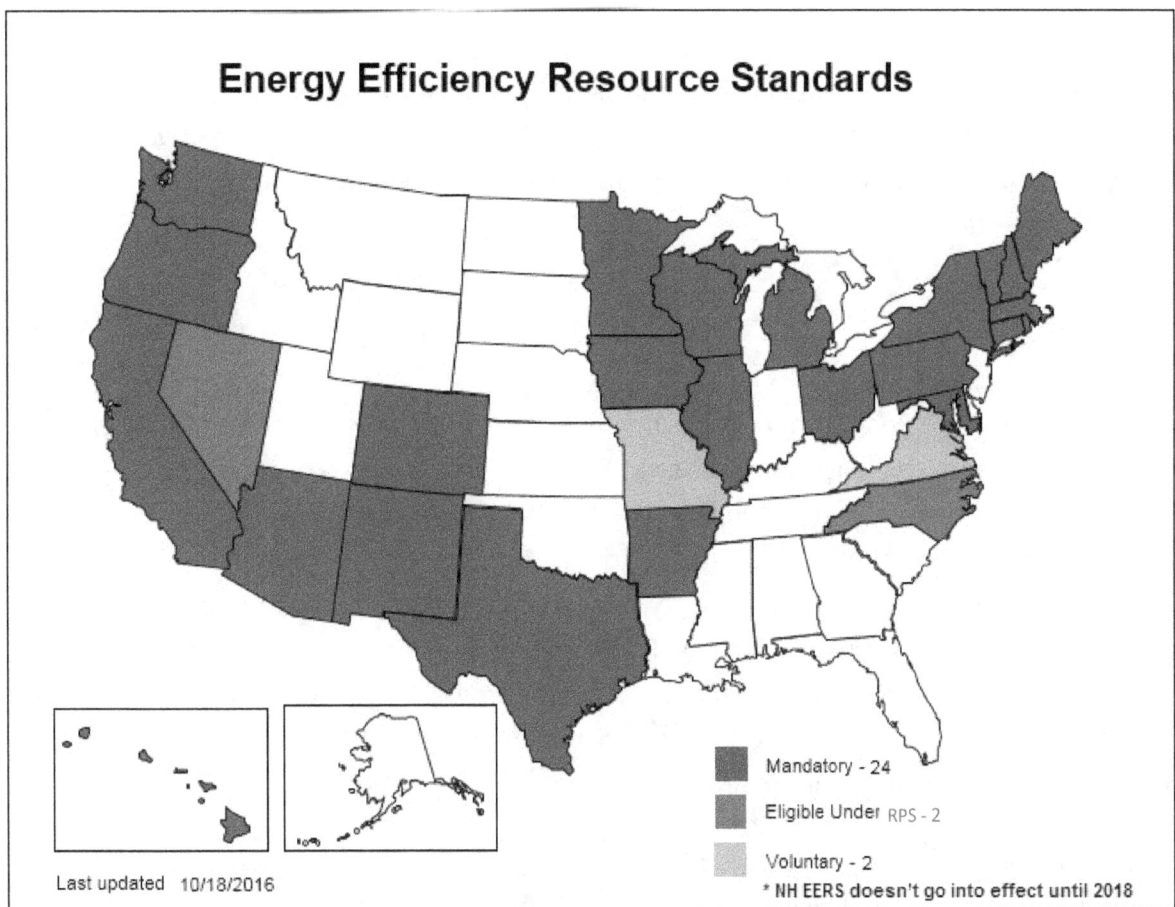

Most states are meeting or on track to meet their incremental savings goals, which typically range from an annual reduction in electricity of about 0.1–2.5 percent.[92] In 2014, incremental savings across the 50 states were equivalent to 0.69 percent of retail electricity sales.[93] In 2012,

---

[90] See footnotes 81 and 82.

[91] States with voluntary EERS: Virginia and Missouri. States eligible under RPS: Nevada, North Carolina. For Nevada, energy efficiency may meet a quarter of the standard through 2014, but is phased out of the RPS by 2025. For North Carolina, its Renewable Energy and Energy Efficiency Portfolio Standard requires renewable generation and/or energy savings of 6 percent by 2015, 10 percent by 2018, and 12.5 percent by 2021 and thereafter. Energy efficiency is capped at 25 percent of target, increasing to 40 percent in 2021 and thereafter. Information from: http://aceee.org/files/pdf/policy-brief/eers-04-2014.pdf.

[92] See footnotes 81 and 82.

[93] "The 2015 State Energy Efficiency Scorecard" (American Council for an Energy-Efficient Economy, October 2015; uses data from 2014). Available at: http://aceee.org/research-report/u1509.

15 of 26 states achieved 100 percent or more of their goals, six states met over 90 percent of their goals, five states achieved over 80 percent of their goals, and only one state realized savings below 80 percent of its goal.[94]

ii. Demand-side Energy Efficiency Programs

Description

Demand-side energy efficiency programs are programs designed to advance energy efficiency improvements within a state or utility service area. They are typically implemented to help meet state policies, standards, or objectives such as EERS programs, "all cost effective" energy efficiency goals, integrated resource planning, and other demand-side management program and budget processes.

Policy Mechanics

Design

Demand-side energy efficiency programs include financial incentives to use energy efficient products, make energy efficiency upgrades to improve the performance of residential, commercial, and industrial buildings, and provide technical assistance and information programs to address market and information barriers. Funding for these programs typically comes from charges added to customer utility bills and from revenues raised through emissions allowance auctions, such as under RGGI. The RGGI auction proceeds go to a variety of sources with the authority to run demand-side energy efficiency programs, including those also funded via independent trusts, DOE's Weatherization Assistance Program (WAP), and state-run energy efficiency grant programs for municipalities.[95]

States are also funding energy efficiency programs using revenues from "forward capacity markets" operated by regional electricity operators. Forward capacity markets allow energy suppliers to bid against each other for the amount of capacity they can supply into the electricity market in a future year. Demand-side management programs have been allowed to bid into these markets as an energy source, demonstrating that energy efficiency programs can compete with more traditional forms of electricity supply in meeting the needs of the power grid.

---

[94] Annie Downs and Celia Cui. "Energy Efficiency Resource Standards: A New Progress Report on State Experience." *American Council for an Energy Efficient Economy* (April 2014). Available at: http://aceee.org/research-report/u1403.

[95] RGGI Inc., Investment of RGGI Proceeds Through 2013 (Regional Greenhouse Gas Initiative Inc., April 2015). Available at: http://www.rggi.org/docs/ProceedsReport/Investment-RGGI-Proceeds-Through-2013.pdf.

Authority

Demand-side programs that are a part of EERS programs are typically established through legislation or PUC authority. Other demand-side management programs can arise as a result of utility planning processes and state and local government efforts to ensure all cost-effective energy efficiency and other policy goals are met.

Obligated Parties

Energy efficiency programs can be administered by investor-owned, municipal, or cooperative utilities; third party administrators; or state and local government agencies.

Measurement and Verification

PUCs generally oversee demand-side energy efficiency programs. Program administrators typically rely on independent evaluators to perform EM&V activities that estimate the incremental annual and cumulative energy savings attributable to the programs. These estimates are typically the basis for annual performance reports submitted by retail electricity suppliers or third party administrators to the PUCs. In the case of state and local government agency run programs that are not overseen by the PUC, energy savings are typically estimated to assure proper use of grants or other funds.

Penalties for Noncompliance

As discussed above, some states with an EERS levy direct fines for missing energy efficiency targets or failure to submit an energy efficiency plan. For some programs under PUC oversight, failure to reach certain performance levels may result in an inability to receive an incentive payment or recover all incurred costs. Demand-side programs funded by RGGI proceeds or grants typically do not have penalties for noncompliance. However, state agencies play a role in evaluating these programs and deciding whether funding should continue to flow to them.

Implementation Status

Well-established state demand-side energy efficiency programs have demonstrated their ability to reduce electricity demand.[96] For example, data reported to the U.S. Energy Information Administration (EIA) show that in 2014, incremental annual savings[97] in electricity consumption through demand-side efficiency programs were 268 GWh in Rhode Island, 1,201 GWh in

---

[96] "The Future of Utility Customer-Funded Energy Efficiency Programs in the United States" (Lawrence Berkeley National Laboratory, January 2013). Available at: https://emp.lbl.gov/sites/all/files/lbnl-5803e.pdf.
[97] EIA defines incremental annual savings for a given year as annualized savings caused by new program participants to existing energy efficiency programs, or program participants to new energy efficiency programs.

Arizona, and 599 GWh in Iowa.[98] These reductions are equivalent to 3.5 percent, 1.6 percent, and 1.3 percent of total 2014 retail electricity sales in those states, respectively.[99] According to data and analyses from sources including Lawrence Berkeley National Lab (LBNL), the DOE Energy Information Administration, and the American Council for an Energy Efficient Economy (ACEEE), as well as the EPA's own analysis for the Clean Power Plan, at least ten leading states have either achieved—or have established requirements that will lead them to achieve—annual incremental savings rates of at least 1.5 percent of the electricity consumption that would otherwise have occurred.[100]

In 2014, utilities and administrators in all 50 states and the District of Columbia implemented electricity demand-side energy efficiency programs, and savings from these programs are increasing. State demand-side energy efficiency programs are estimated to have reduced electricity demand by 25.7 million MWh in 2014, or 0.7 percent of national retail electricity sales. These savings are an increase of 5.8 percent over the previous year.[101]

iii. Building Energy Codes

Description

Building energy codes establish minimum efficiency requirements for new and renovated residential and commercial buildings. These measures are intended to eliminate inefficient technologies with minimal impact on up-front project costs. This can reduce the need for energy generation capacity and new infrastructure while reducing energy bills. Energy codes lock in future energy savings during the building design and construction phase, rather than through a renovation.

Policy Mechanics

Design

Codes specify "thermal resistance" improvements to the building shell and windows, minimum air leakage, and minimum efficiency for heating and cooling equipment.

Based on provisions of the Energy Policy Act of 1992, the International Energy Conservation Code (IECC) is the prevailing national model code for residential buildings. Similarly, American

---

[98] "Electric Power Sales, Revenue, and Energy Efficiency Form EIA-861 Detailed Data Files" (Energy Information Administration, January 2016). Available at: http://www.eia.gov/electricity/data/eia861/.
[99] "Electricity: Detailed State Data" (Energy Information Administration, October 2015). Available at: http://www.eia.gov/electricity/data/state/.
[100] See EPA's Demand-Side Energy Efficiency Technical Support Document (August 2015) for more information. Available at: https://www.epa.gov/sites/production/files/2015-11/documents/tsd-cpp-demand-side-ee.pdf.
[101] "The 2015 State Energy Efficiency Scorecard" (American Council for an Energy-Efficient Economy, October 2015; uses data from 2014). Available at: http://aceee.org/research-report/u1509.

Society of Heating, Refrigerating and Air-Conditioning Engineers (ASHRAE) Standard 90.1 serves as the national model commercial code. The most current codes are ASHRAE Standard 90.1-2013 and 2015 IECC.[102]

Building code standards are revised every three years. The IECC codes are updated every 18 months using a prescribed process, and new editions are published every three years. The ASHRAE Standard 90.1 revision process occurs on a three-year cycle, but proposals for revisions are accepted at any time.[103] The Energy Policy Act of 1992 requires DOE to conduct determinations on each successive version of the IECC residential code provisions and ASHRAE Standard 90.1. For residential buildings, each state must consider adoption of each successive version of the IECC for which DOE makes a positive determination on energy savings, and report to DOE within two years on whether they have adopted the new version. For commercial buildings, state adoption of each successive version of ASHRAE Standard 90.1 is mandatory subject to the same DOE determination process; however, there are no penalties for states that do not comply.[104]

By locking in efficiency measures at the time of construction, codes are intended to capture energy savings that are more cost-effective than retrofit opportunities available after a building has been constructed. Energy code requirements are also intended to overcome market barriers to efficient construction in both the commercial and residential sectors, such as the complexity of advanced codes, lack of local-level implementation resources, and a shortage of empirical data on the costs and benefits of codes.

Authority

Model building codes are typically developed at the national or international level, adopted at the state and/or local level, and implemented and enforced locally.

Obligated Parties

Local parties, such as developers and property owners requiring building permits, are the most common obligated parties.

---

[102] "Code Adoption Status: February 2016," Building Codes Assistance Project. Available at: http://bcap-energy.org/wp-content/uploads/2015/11/code_status_february_20161.pdf.

[103] "Code Development," Online Code Environment & Advocacy Network, accessed March 11, 2016. Available at: http://energycodesocean.org/research-topic/code-development#.

[104] "Regulations & Rulemaking," Building Energy Codes Program, U.S. Department of Energy, Energy Efficiency & Renewable Energy, accessed March 11, 2016. Available at: https://www.energycodes.gov/regulations.

Measurement and Verification

Program implementation steps, including builder training, compliance assurance, and enforcement, are typically the responsibility of state and local governments. These steps, however, are often not fully or uniformly implemented for numerous reasons, including an emphasis on health and safety issues over the proper functioning of mechanical equipment, a lack of trained staff to review building plans and conduct onsite inspections, and limited funding to carry out key implementation activities. As a result, most jurisdictions do not have the capacity to analyze code compliance and to identify the measures and strategies that should be targeted for improved implementation.

Penalties for Noncompliance

In order to get building permits approved, the relevant developer or property owners must show they are in compliance with standards. Since permitting is done at the local level, the use of penalties and the ability to enforce standards vary significantly by region. DOE has been working with states and localities to improve compliance practice.

Implementation Status

As of September 2015, 25 states have adopted IECC 2009 residential energy codes, 10 states and Washington, D.C., have adopted the IECC 2012, while two states have gone further by adopting the IECC 2015. In the commercial sector, 21 states have adopted ASHRAE 90.1-2007, 18 states and Washington, D.C., have adopted ASHRAE 90.1-2010, and two states have adopted ASHRAE 90.1-2013. Currently, 12 states have outdated or no statewide residential energy code, and 11 states have outdated or no statewide energy codes for commercial construction.[105] The current status of state residential and commercial energy codes are shown below in Figure 6 and Figure 7, respectively.[106] Illinois is notable as a state that adopted the 2012 IECC on January 1, 2013, and has set up an aggressive system for implementing future updates to energy building codes. A provision in past legislation to adopt 2009 IECC and ASHRAE 90.1-2007 directed the state's Capital Development Board to adopt subsequent versions of the IECC within 9 months of publication. DOE expects Illinois's energy cost savings to reach $270 million annually by 2030.[107]

---

[105] Building Codes Assistance Project, Code Adoption Status, September 2015. Available at: http://energycodesocean.org/sites/default/files/code%20status%201%20pgr%20new.pdf.

[106] "Code Development," Online Code Environment & Advocacy Network, accessed March 11, 2016. Available at: http://energycodesocean.org/research-topic/code-development#.

[107] U.S. EPA, *Clean Energy-Environment Guide to Action* (U.S. Environmental Protection Agency, 2015), accessed March 10, 2016. Available at: http://epa.gov/statelocalclimate/resources/action-guide.html.

Figure 6: Residential State Energy Code Status

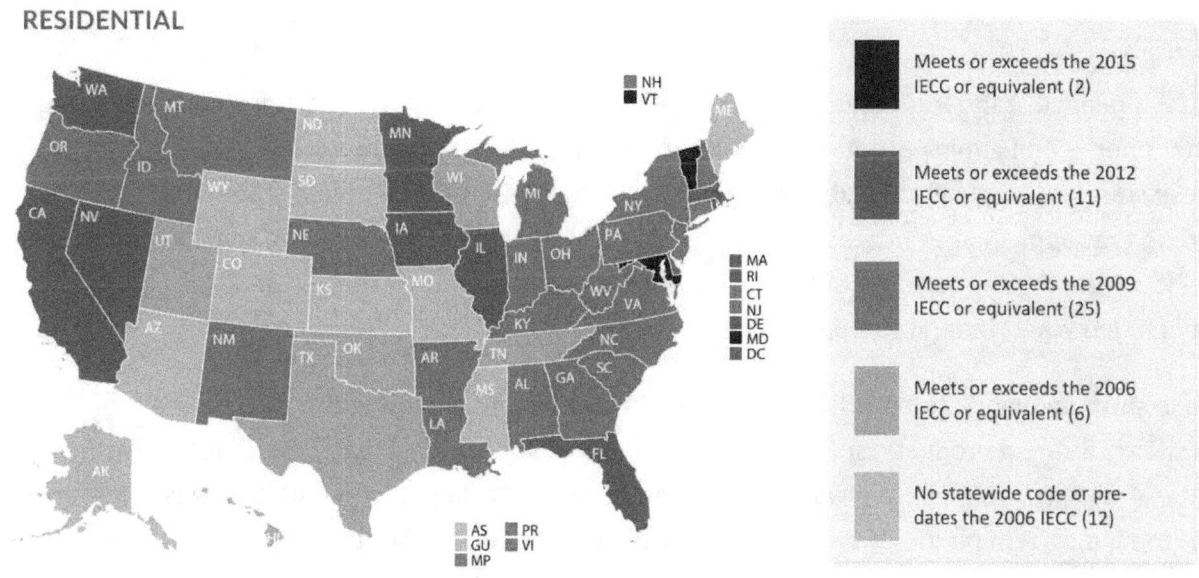

Source: Building Codes Assistance Project. Code Adoption Status (September 2015). Available at:
http://energycodesocean.org/sites/default/files/code%20status%201%20pgr%20new.pdf.

Figure 7: Commercial State Energy Code Status

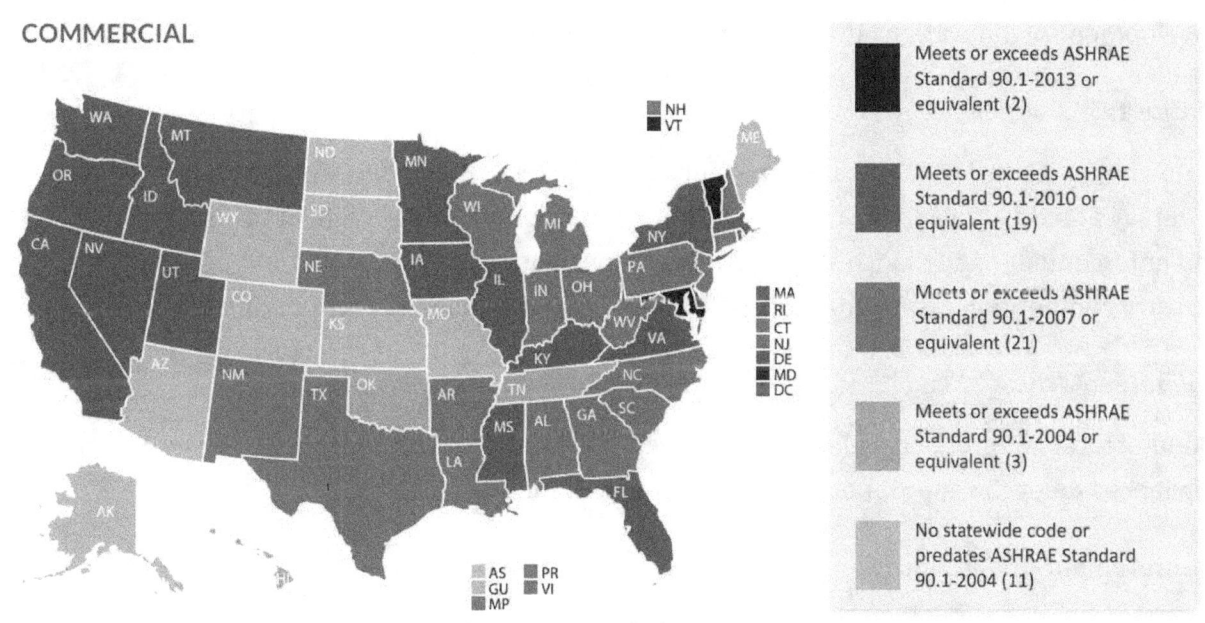

Source: Building Codes Assistance Project. Code Adoption Status (September 2015). Available at:
http://energycodesocean.org/sites/default/files/code%20status%201%20pgr%20new.pdf.

iv. Appliance and Equipment Efficiency Standards

Description

State appliance standards establish minimum energy-efficiency levels for those appliances and other energy-consuming products that are not already covered by the federal government. These standards typically prohibit the sale of less efficient models within a state. States are finding that appliance standards offer a cost-effective strategy for improving energy efficiency and lowering energy costs for businesses and consumers, although these standards are superseded when federal standards are enacted for new product categories.

While state appliance standards can be useful in testing and exploring the effectiveness of standards for new products, states cannot preempt or supersede existing federal standards. States may apply to DOE for a waiver to implement more stringent standards. This is sometimes granted if a certain period of time has passed since the federal standard has been updated.

Policy Mechanics

Design

When states implement appliance and equipment standards, they are establishing a minimum efficiency for products, such as refrigerators or air conditioners, thereby reducing the energy associated with using the product. Standards prohibit the production and sale of products less efficient than the minimum requirements, encouraging manufacturers to focus on how to incorporate energy-efficient technologies into their products at the least cost and hastening the development of innovations that bring improved performance.

Authority

State energy offices, which typically administer the federal state energy program funds, have generally acted as the administrative lead for standards implementation. In contrast, inspection and enforcement of appliance standards regulations has typically involved self-policing. Industry competition is such that competitive manufacturers usually report violations.

Obligated Parties

Manufacturers of products being sold in a given state are typically obligated to ensure their appliances meet the appropriate energy efficiency standards.

Measurement and Verification

Evaluating the benefits and costs of the standards is important during the standards-setting process. Once enacted, however, little field evaluation is performed.

Penalties for Noncompliance

Appliances and equipment found in violation of the minimum energy performance standards are not allowed to be sold or manufactured in the state.

Implementation Status

As of March 2016, 16 states and Washington, D.C., have enacted appliance efficiency standards since 2001. However, most of these standards have been superseded by federal standards. Still, 11 states (AZ, CA, CO, CT, NH, MD, OR, GA, TX, RI, WA) and Washington, D.C., have either enacted standards for equipment not covered federally or obtained waivers to enact tougher appliance standards where the federal regulations have become outdated. California currently leads all states in active state standards, covering 17 products, including pool pumps and hot tubs, vending machines, televisions, battery chargers, toilets and urinals, bottle-type water dispensers, faucets, CD and DVD equipment, and various lighting applications.[108]

v. Incentives and Finance Mechanisms for Energy Efficiency

Description

States offer a diverse portfolio of financing and incentive approaches that are designed to address specific financing challenges and barriers and incentivize specific markets and customer groups to invest in energy efficiency. These programs include revolving loan funds, energy performance contracting, green banks, tax incentives, rebates, grants, and other incentives.

Policy Mechanics

Design

Revolving loan funds provide low-interest loans for energy efficiency improvements. The funds are designed to be self-supporting. States create a pool of capital that "revolves" over a multi-year period, as payments from borrowers are returned to the capital pool and are subsequently lent to other borrowers. Revolving loan funds can be created from several sources, including public benefits funds (PBFs),[109] utility program funds, general state revenues, or federal funding sources. Revolving funds can grow in size over time, depending on repayment interest rates and program administrative costs.

---

[108] Appliance Standards Awareness Project, State Standards, accessed March 10, 2016. Available at: http://www.appliance-standards.org/states.

[109] PBFs are dedicated funds used for supporting research and development of energy efficiency and renewable energy projects. Funds are normally collected either through a small charge for every electric customer or through specified contributions from utilities.

Energy performance contracting allows the public sector to contract with private energy service companies (ESCOs) to provide building owners with energy-related efficiency improvements that are guaranteed to save more than they cost over the course of the contracting period. ESCOs provide energy auditing, engineering design, general contracting, and installation services, and help arrange project financing.[110] The contracts are privately funded and do not involve state funding or financial incentives.

Green banks offer an emerging approach used by an increasing number of states to evolve away from traditional state funded incentive programs. They use creative financing to bring and leverage private capital to develop projects and markets. Green banks can be self-sufficient and manage their seed capital in perpetuity. They do not require ongoing funding from the legislative and state budget process once they are capitalized. Because green banks are effectively nonprofit organizations, they can offer a capital cost far lower than any other source of capital available in the market. States can consolidate their existing incentive programs and resources under a green bank framework.[111]

State tax incentives for energy efficiency are available as personal or corporate income tax credits, tax exemptions (e.g., sales tax exemptions on energy-efficient appliances), and tax deductions (e.g., for construction programs). Tax incentives aim to spur private sector innovation to develop more energy efficient technologies and practices and increase consumer choice of energy-efficient products.[112]

Rebates (also known as "buy-downs") are used to promote demand-side energy efficiency reductions by providing direct incentives to customers who purchase or make upgrades to approved efficient appliances or retrofit their homes (e.g., a utility may refund part of the cost for a homeowner to improve attic insulation or purchase a high-efficiency furnace). Funding for rebates may come from PBFs, direct grants, or utility program funds.

Grants from the federal government, state government, regional agency, or private source may be used to start or finance energy efficiency programs. A grant may be used to provide funding for a specific construction project (e.g., retrofit of a school), finance a rebate program, initiate a revolving fund, conduct a behavior change campaign (e.g., educate public about the benefits of off-peak energy use), or any other type of program that meets the specific grant requirements.

---

[110] U.S. EPA, *Integrating State and Local Environmental and Energy Goals: Energy Performance Contracting - Fact Sheet* (U.S. Environmental Protection Agency, September 2004).

[111] U.S. EPA, *Clean Energy-Environment Guide to Action* (U.S. Environmental Protection Agency, 2015), accessed March 10, 2016. Available at: http://epa.gov/statelocalclimate/resources/action-guide.html.

[112] Elizabeth Brown, Harvey Sachs, Patrick Quinlan, and Daniel Williams. "Tax Credits for Energy Efficiency and Green Buildings: Opportunities for State Action." *American Council for an Energy Efficient Economy* (2002).

Authority

Financial mechanisms and incentives for energy efficiency are run by utilities and state and local governments. Utilities primarily offer rebates, grants, and loans. Personal, corporate, sales, and property tax incentives are mainly offered by state and local governments.[113]

Implementation Status

Financial mechanisms and incentives for energy efficiency exist in all 50 states, with the most prevalent financial mechanisms and incentives for energy efficiency being rebates and loan programs in place. There are 50 tax incentives and over a thousand rebate, grant, and loan programs that help finance and deliver electricity savings.[114] Texas LoanSTAR, also known as the Loans to Save Taxes and Resources program, began in 1988 as a $98.6 million retrofit program for energy efficiency in buildings (primarily public buildings such as state agencies, local governments, and school districts). As of January 2014, the program has funded over 237 loans, totaling more than $395 million. The program has also saved over $419 million and reduced CO_2 emissions by 3.7 million tons.[115]

C. Renewable Energy Policies and Programs

States have adopted a range of requirements and programs to advance the deployment of renewable energy technologies, including renewable portfolio standards, performance-based incentives, and public benefit funds.[116] These renewable energy policies and programs reduce GHG emissions by increasing the use of renewable energy and altering the mix of energy supply.

i. Renewable Portfolio Standards

Description

A renewable portfolio standard (RPS), also known as a renewable electricity standard (RES), is a mandatory requirement for retail electricity suppliers to supply a minimum percentage or amount of their retail electricity load with electricity generated from eligible sources of

---

[113] "Programs," Database of State Incentives for Renewables & Efficiency, accessed March 10, 2016. Available at: http://programs.dsireusa.org/system/program.

[114] Ibid.

[115] Texas State Energy Conservation Office, "LoanSTAR Revolving Loan Program," accessed March 24, 2016. Available at: http://www.seco.cpa.state.tx.us/ls/.

[116] Feed-in tariffs, a performance-based incentive, offer long-term purchase agreements to renewable energy electricity generators. Public benefit funds are typically created by levying a small fee as a part of retail electricity rates and are used to support rebate, loan, and other programs that support renewable energy deployment. For more information, see Database of State Incentives for Renewables and Efficiency. Available at: http://www.dsireusa.org/.

renewable energy.[117] An RPS indirectly affects EGU CO_2 emissions by reducing the utilization of fossil fuel–fired EGUs. As of March 2016, 29 states and Washington, D.C., have adopted a mandatory RPS (see Figure 8), although designs vary (e.g., applicability, targets and timetables, geographic and resource eligibility, alternative compliance payments) and an additional eight states have voluntary renewable goals.[118,119]

Figure 8: States with Renewable Portfolio Standards

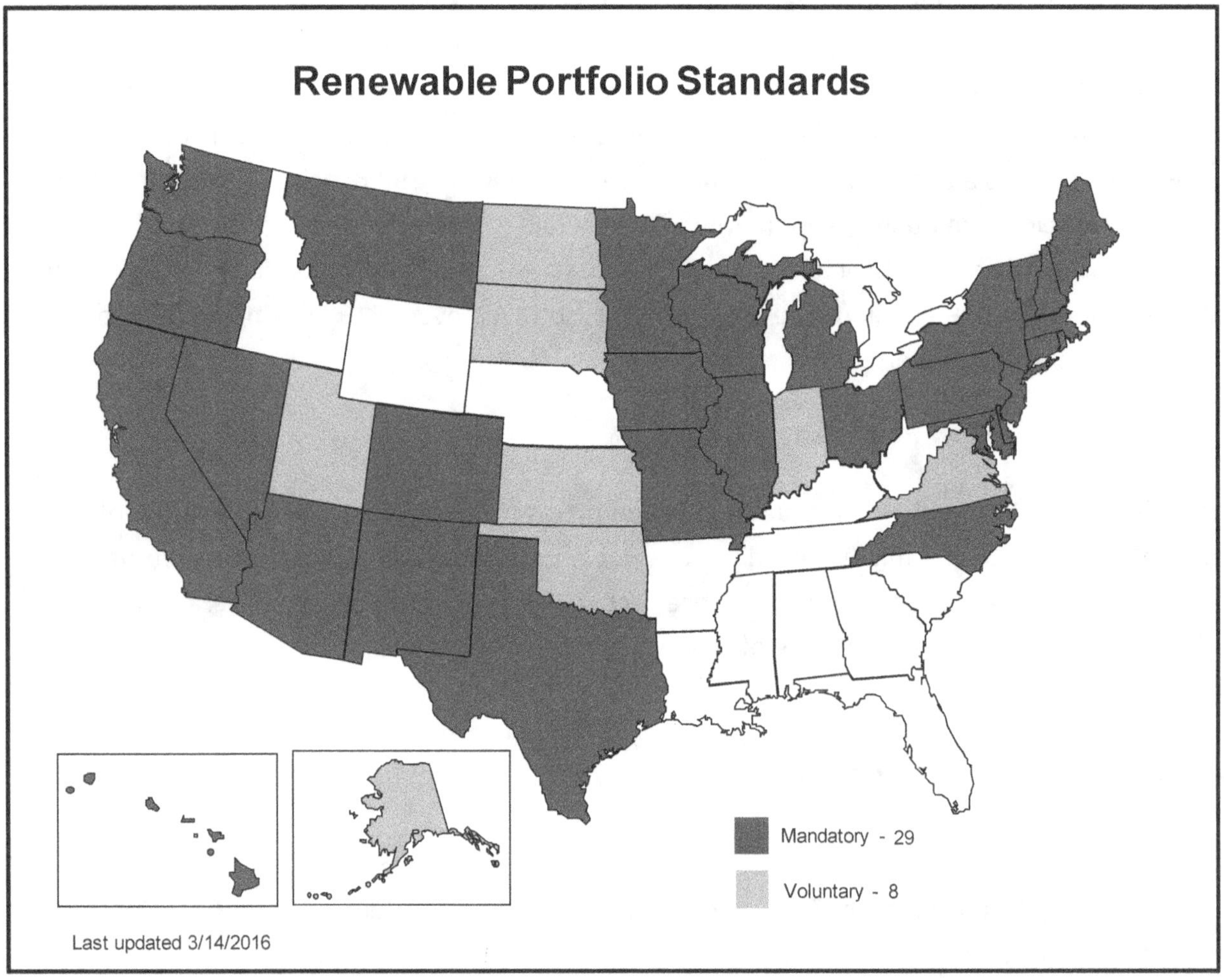

[117] In some state Renewable Portfolio Standards (alternatively called "Alternative and Renewable Energy Portfolio Standards"), selected non-renewable sources such as coal bed methane or gasification are eligible for credit.
[118] Database of State Incentives for Renewables and Efficiency (June 2015), accessed March 10, 2016. Available at: http://www.dsireusa.org/.
[119] Alaska House Bill 306, Signed by Governor Sean Parnell June 16, 2010. Available at: http://www.legis.state.ak.us/basis/get_bill_text.asp?hsid=HB0306Z&session=26.

Design

RPS requirements typically start at modest levels and ramp up over a period of several years. An RPS relies on market mechanisms to increase electricity generation from eligible sources of renewable energy.

Retail electricity suppliers can comply with RPS requirements through several mechanisms, which vary by state, including:

- Ownership of a qualifying renewable energy facility and its electric generation output.
- Purchasing electricity bundled with renewable energy certificates (RECs)[120] from a qualifying renewable energy facility.
- Purchasing RECs separately from electricity generators. Unlike bundled renewable energy, which is dependent on physical delivery via the power grid, RECs can be traded between any two parties, regardless of their location. However, state RPS rules typically condition the use of RECs based on either location of the associated generation facility or whether it sells power into the state or to the regional grid.

Authority

Most state RPSs are established through legislation and administered by state PUCs.

Obligated Parties

RPS applicability varies by state. All state RPSs apply to investor-owned utilities, while some state RPS obligate municipal utilities, rural cooperatives, and/or other retail providers, often depending on a minimum number of customers served.

Measurement and Verification

Some state RPSs include an alternative compliance payment (ACP) option, where a retail electricity supplier may purchase compliance credits from the state at a known price, which acts as a de facto price cap, if it has not procured sufficient electricity from renewable energy sources or RECs to meet the RPS compliance requirement. State PUCs typically require annual compliance reports from retail electricity suppliers subject to a RPS. Most states use regional tracking systems (e.g., Western Renewable Energy Generation Information System, PJM

---

[120] RECs represent the non-energy attributes, including all the environmental attributes, of electricity generation from renewable energy sources. RECs are typically issued in single MWh increments.

Generation Attribute Tracking System) to issue, track, and retire RECs for RPS compliance purposes.[121]

Penalties for Noncompliance

States have developed a range of compliance enforcement and flexibility mechanisms. As of 2007, despite the fact that several states had not achieved the RPS targets, only Connecticut and Texas had levied fines. A $5.6 million penalty was incurred in Connecticut in 2006. In 2003 and 2005, two competitive electricity service providers in Texas were penalized a total of $4,000 and $28,000 respectively.[122] More recently the vast majority of states have met their RPS requirements, and for those that have not, utilities have been allowed to "make-up" shortfalls in subsequent years.[123]

ACPs that are recycled to support other renewable and efficiency measures have helped other states avoid penalties for noncompliance.[124] The reported compliance cost for the entire RPS in the District of Columbia was $2.6 million in 2011, of which ACPs made up $229,500. Electricity suppliers in Maryland submitted more than 4.6 million RECs for compliance in 2011 for a total cost of $14.6 million, of which $98,520 came from ACPs.[125]

Implementation Status

States with RPS policies have demonstrated higher levels of renewable energy capacity development. From 1998 to 2012, 67 percent (46 GW) of all non-hydro renewable capacity additions occurred in states with active or impending RPS requirements, although other factors may contribute to the growth in renewable capacity.[126]

---

[121] For a summary of REC tracking systems, see: U.S. Department of Energy Renewable Energy Certificates, National REC Tracking Systems, accessed March 10, 2016. Available at: http://apps3.eere.energy.gov/greenpower/markets/certificates.shtml?page=3.

[122] Ryan Wiser and Galen Barbose, _Renewables Portfolio Standards in the United States – A Status Report with Data Through 2007_ (Lawrence Berkeley National Laboratory, April 2008). Available at: http://emp.lbl.gov/sites/all/files/REPORT%20lbnl-154e-revised.pdf.

[123] Personal communication with Galen Barbrose of Lawrence Berkeley National Lab, March, 2016.

[124] Ryan Wiser and Galen Barbose, _Renewables Portfolio Standards in the United States – A Status Report with Data Through 2007_ (Lawrence Berkeley National Laboratory, April 2008). Available at: http://emp.lbl.gov/sites/all/files/REPORT%20lbnl-154e-revised.pdf.

[125] J. Heeter, G. Barbose, L. Bird, S. Weaver, F. Flores-Espino, K. Kuskova-Burns, and R. Wiser, "A Survey of State-Level Cost and Benefit Estimates of Renewable Portfolio Standards" (NREL, May 2014). Available at: http://www.nrel.gov/docs/fy14osti/61042.pdf.

[126] Galen Barbose, _Renewables Portfolio Standards in the United States: A Status Update_ (Lawrence Berkeley National Laboratory, November 2013). Available at: http://emp.lbl.gov/sites/all/files/rps_summit_nov_2013.pdf.

ii. Performance-based Incentives and Finance Mechanisms for Renewable Energy

Description

States offer a diverse portfolio of financing mechanisms, performance-based incentives, and state utility ratemaking approaches that are designed to address specific financial challenges and barriers, and help specific markets and customer groups produce clean energy.

Policy Mechanics

Design

States support the advancement of clean generation technologies through performance-based incentives, including feed-in tariffs and other payments, or tax incentives. Performance-based incentives are paid based on the actual energy production of a system. Feed-in tariffs establish temporarily elevated price per kWh in order to encourage renewable energy innovation using high cost technologies. Tax incentives are used to lower financial barriers to renewable energy production.

A major source of funding for renewable energy activities comes from PBFs, but states also fund these activities through alternative sources including direct grants, rebates and generation incentives provided by utilities.

State tax incentives for renewable energy and combined heat and power (CHP) take the form of personal or corporate income tax credits and tax exemptions. State tax incentives for renewable energy are a common policy tool, mainly using credits on personal or corporate income tax and exemptions from sales tax, excise tax, and property tax.

Authority

Financial mechanisms and incentives for renewables are run by utilities, non-profits, and state and local government. Personal, corporate, sales, and property tax incentives are mainly offered by state and local government.[127]

Implementation Status

Financial mechanisms and incentives for renewable energy of some form exist in most states. According to the Database of State Incentives for Renewable Energy (DSIRE), there are more

---

[127] "Summary Tables" (Database of State Incentives for Renewables & Efficiency), accessed March 10, 2016. Available at: http://programs.dsireusa.org/system/program/tables.

than 200 tax incentives. In addition, over 50 performance-based incentives are offered from state and local governments, as well as utilities and non-profits.[128]

There are currently 26 states that offer some form of performance-based incentive, and in several other states utilities have adopted programs based on performance-based incentives, including feed-in tariffs, standard offer payments, and payments in exchange for RECs.[129] In many cases, however, PBI is limited to customer-sited projects or limited by size eligibility.

Financial incentives, working in concert with a strong RPS and net metering policies,[130] have contributed to the rapid growth in solar power deployment in New Jersey. The state's RPS includes a minimum carve-out for solar sources, and allows solar energy generators to earn solar renewable energy certificates (SRECs) that can then be sold to electricity suppliers trying to meet the minimum solar production and/or purchase requirement. As a result of these interdependent policies, the number of solar photovoltaic facilities grew, with total capacity in New Jersey increasing by 20 percent from 2013 to 2014.[131] New Jersey ranks second only to California in terms of total installed PV capacity.[132]

D. Utility Planning Approaches and Requirements

Description

Some public utility commissions require utilities to conduct portfolio management or integrated resource planning (IRP) to ensure the supply of least cost and stable electric service to customers over the long term. Portfolio management refers to energy resource planning that incorporates a variety of energy resources, including supply-side (e.g., traditional and renewable energy sources) and demand-side (e.g., energy efficiency) options. The term "portfolio management" typically describes resource planning and procurement in states that have restructured their electric industry and may be required for default service providers (the backup electric service provider in areas open to competition). IRP is generally used by vertically integrated utilities and is a long-range planning process to meet forecasted demand for energy within a defined geographic area through a combination of supply-side resources and demand-side resources and considering a broad range of perspectives. The goal of an IRP is

---

[128] "Summary Tables" (Database of State Incentives for Renewables & Efficiency), accessed March 10, 2016. Available at: http://programs.dsireusa.org/system/program/tables.

[129] "Summary Tables" (Database of State Incentives for Renewables & Efficiency), accessed March 10, 2016. Available at: http://programs.dsireusa.org/system/program?type=13&.

[130] Net metering policies allow solar installation owners to receive a credit on their utility bill for the excess electricity generated by solar panels that is fed back into the grid.

[131] Solar Energy Industries Association, "New Jersey Ranks 3rd in U.S. in Total Solar Capacity" (*seia.org*, March 17, 2015). Available at: http://www.seia.org/news/new-jersey-ranks-3rd-us-total-solar-capacity.

[132] "Open PV State Rankings," National Renewable Energy Laboratory, accessed March 10, 2016. Available at: https://openpv.nrel.gov/rankings.

to identify the mix of resources that will minimize future energy system costs while ensuring safe and reliable operation of the system.

In addition to energy resource planning, two states have policies or requirements for utilities to specifically factor pollution reduction requirements into their planning. In Colorado, the Clean Air Clean Jobs Act (CACJA), signed into law on April 19, 2010, requires utilities to submit a plan to the PUC showing how they would meet EPA standards for a variety of pollutants.[133] The law was passed because the state was out of compliance with the national Ambient Air Quality Standard for Ozone, and the EPA threatened to propose more stringent standards for the state.

In 2001, Minnesota enacted Minnesota Statute 216B.1692, which encourages utilities to make voluntary emissions reductions and provides them with a mechanism to recover the costs through customer rate increases outside of the normal rate review cycle.[134]

Policy Mechanics

Design

- Portfolio Management and IRP – Portfolio management emphasizes diversity in fuels, technologies, and power supply contract durations. Portfolio management includes energy efficiency and renewable generation as key strategic components. Portfolio management typically involves a multi-step process of forecasting, resource identification, scenario analysis, and resource procurement.

  Several states and vertically integrated utilities rely on an IRP process for long-term planning. Since these utilities own generation assets, they use their IRPs to evaluate a broad range of options for meeting electricity demand over a 20- or 30-year time frame. The IRP considers new supply-side options (including renewable resources) and demand-side options, and purchased power (including transmission considerations). A broad range of plans are considered, reflecting a range of objectives and capturing key uncertainties. Plans are evaluated against established criteria (e.g., costs, rate impacts, emissions, diversity, etc.) and are ranked. The IRPs detail fuel and electricity price information, customer demand forecasts, existing plant performance, other plant additions in the region, and legislative decisions. The following examples show how various states have designed their programs:

---

[133] Regulatory Assistance Project, *Addressing the Effects of Environmental Regulations: Market Factors, Integrated Analyses, and Administrative Processes* (RAP, 2013). Available at: www.raponline.org/document/download/id/6455.

[134] Minnesota PUC, *Report To The Legislature On Emissions Reduction Projects Under Minnesota Statutes 216B.1692* (Minnesota Public Utilities Commission, March 2008). Available at: http://mn.gov/puc/documents/pdf_files/000661.pdf.

- Montana is a deregulated state that has established least cost planning rules and policy guidelines for default electricity suppliers. These rules and guidelines target long-term electricity supply and are slightly different for vertically integrated utilities and restructured utilities. Vertically integrated utilities are required to submit electric supply resource plans every two years with the aim of providing a balanced, environmentally responsible electricity portfolio. Meanwhile, restructured utilities must file updates to their portfolio action plans every three years.[135] These plans must include supply-side and demand-side resources, and they must address the need to supply power in a way that minimizes the environmental cost by estimating the cost to the environment of alternatives. In addition, utilities must account for the costs of complying with existing and future environmental regulations. When considering various resource options, Montana requires a competitive solicitation process, allowing resource operators and developers to submit their proposals to the default electricity supplier for consideration. Montana also requires the portfolio management plans to be subject to an advisory committee review and a public review.[136]

- Oregon electric utilities submit IRPs every two years, covering a 20-year timeframe. The goal of these plans is to consider the acquisition of resources at least cost while keeping the public interest in mind. Potential risk factors must be considered, including price volatility, weather, and the cost of meeting existing and future federal environmental regulations. Quantifiable environmental externalities are included, as are less quantifiable developments such as changes in market structure and the establishment of a renewable portfolio standard. As for energy efficiency requirements during the planning process, Oregon determines these on a utility-by-utility basis.[137]

- Multi-Pollutant Utility Planning – Two states, Minnesota and Colorado, have worked collaboratively with their investor-owned utilities to develop multi-pollutant emissions reduction plans on a utility-wide basis. This multi-pollutant, collaborative approach

---

[135] Rachel Wilson and Bruce Biewald, *Best Practices in Electric Utility Integrated Resource Planning* (RAP, 2013). Available at: http://www.raponline.org/document/download/id/6608.
[136] U.S. EPA, *Clean Energy-Environment Guide to Action* (U.S. Environmental Protection Agency, 2015), accessed March 10, 2016. Available at: http://epa.gov/statelocalclimate/resources/action-guide.html.
[137] Ibid.

enables utilities to determine the least cost way to meet long-term and comprehensive energy and environmental goals.

- The Colorado CACJA requires investor-owned utilities (IOUs) with coal plants to submit a multi-pollutant plan to the PUC to meet the EPA standards for NO_x, SO_2, particulates, mercury, and CO_2. Utilities were not required to adopt a specific plan set by the state, but had to meet with Colorado Department of Public Health and Environment (CDPHE) and PUC approval. Xcel Energy's plan was submitted and approved in 2010.[138]

- The Minnesota Emissions-Reductions Rider allows utilities to submit plans for projects that reduce emissions and go beyond federal requirements outside of a general rate case. It allows them to recover the costs of those actions as an incentive.[139] The specific design and process of the projects vary by utility, but typically involve installing additional pollution control equipment at coal-fired power plants, or repowering them with natural gas.

Authority

State utility commissioners oversee utilities' and default service providers' procurement practices in their states. Typically, the commissions solicit comments and input as they develop portfolio management practices from a wide variety of stakeholders. The utility regulator may also play a role in reviewing and approving utilities' planning procedures, selection criteria, and/or their competition solicitation processes.

Obligated Parties

Vertically integrated utilities are often obligated under integrated resource planning, while in restructured markets, the default utility service provider may be obligated to conduct portfolio management.

For multi-pollutant planning, Colorado IOUs, Xcel Energy and Black Hills Energy were required to file plans with the Department of Public Health and Environment and the PUC in order to be compliant with the CACJA. Plans needed to meet the National Ambient Air Quality Standards for a number of air pollutants.

---

[138] Regulatory Assistance Project, *Addressing the Effects of Environmental Regulations: Market Factors, Integrated Analyses, and Administrative Processes* (RAP, 2013). Available at: www.raponline.org/document/download/id/6455.

[139] Minnesota Office of Revisor of Statutes, 2013 Minnesota Statutes, §216B,1692 Emissions-Reduction Rider , 2013, accessed March 10, 2016. Available at: https://www.revisor.mn.gov/statutes/?id=216B.1692.

As the Minnesota multi-pollutant legislation is voluntary for state utilities, there is neither compliance nor reporting requirements.

Measurement and Verification

Regulatory oversight aims to ensure utilities are following through with their plans. Regulators often require utilities to submit portfolio management plans and progress reports at regular intervals. These plans and reports describe in detail the assumptions used, the opportunities assessed, and the decisions made when developing resource portfolios. Regulators then carefully review these plans and either approve them or reject them and recommend changes needed for approval. California, for example, requires utilities to submit biennial IRPs and quarterly reports on their plans.

Penalties for Noncompliance

There are no penalties for noncompliance, however there is usually significant interaction with the regulator during the planning and implementation process as is described above.

Implementation Status

As of 2015, more than two-thirds of the states have integrated resource or other long-term planning requirements,[140] while Minnesota and Colorado have multi-pollutant planning policies or requirements (see Figure 9).

---

[140] Wilson, Rachel and Bruce Biewald, *Best Practices in Electric Utility Integrated Resource Planning* (RAP, June 2013). Available at: http://www.raponline.org/document/download/id/6608.

Figure 9: States with Integrated Resource Planning or Similar Processes

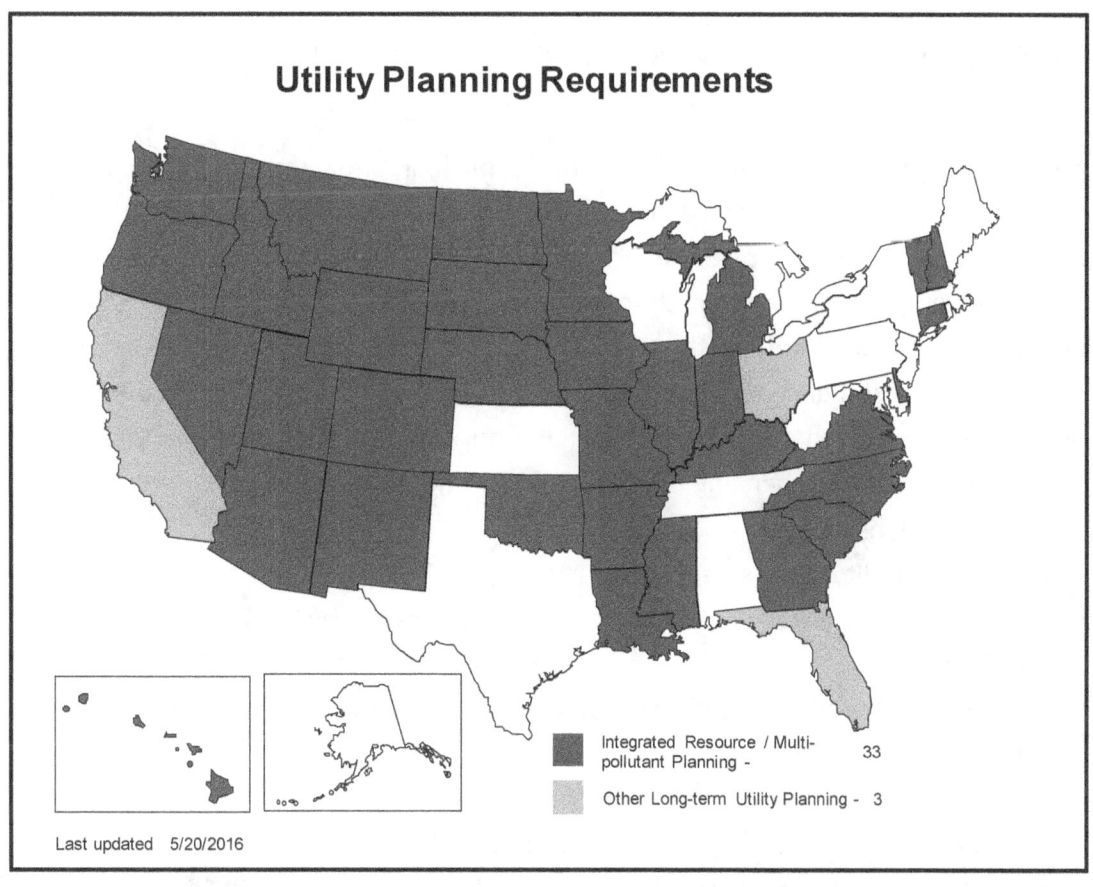

Utility Planning Requirements

Integrated Resource / Multi-pollutant Planning - 33

Other Long-term Utility Planning - 3

Last updated 5/20/2016

Primary source: *Clean Energy-Environment Guide to Action* (U.S. Environmental Protection Agency, 2015), accessed May 20, 2016. Available at: http://epa.gov/statelocalclimate/resources/action-guide.html.[141]

In Missouri, for example, Ameren's 2014 Integrated Resource Plan Update calls for:

- Spending $148 million from 2016-2018 to achieve 426 GWh of energy savings and 114 MW of peak demand savings.
- Installing 400 MW of wind power, 45 MW of solar power, 20 MW of hydroelectric power, and 5 MW of landfill gas capacity by 2034.
- Installing 600 MW of combined-cycle natural gas capacity by 2034.
- Retiring one-third of coal-fired generating capacity by 2034.
- Planning for a 12 percent increase in energy consumption, 8 percent increase in peak demand, and 0.59 percent annual retail sales increase by 2034.
- Incorporating a carbon price of between $23 and $53/ton beginning in 2025.

---

[141] Additional sources for other long-term planning requirements include: California Office of Ratepayer Advocates, *Long Term Procurement Planning: 2014 – 2024*, accessed May 20, 2016. Available at: http://www.ora.ca.gov/ltpp.aspx. Florida Public Service Commission, *Ten-Year Site Plans*, accessed May 20, 2016. Available at: http://www.psc.state.fl.us/ElectricNaturalGas/TenYearSitePlans. LAWriter Ohio Laws and Rules, *4901:5-5-06 Resource plans*, accessed May 20, 2016. Available at: http://codes.ohio.gov/oac/4901%3A5-5-06.

- Planning for a natural gas price increase of between $4-6/MMBtu by 2034.[142]

In Virginia and North Carolina, Dominion Resources filed an updated integrated resource plan in April 2016. Key highlights from the report include:

- Five detailed "Study Plans" (including solar, co-fire, nuclear, and wind). The company's integrated resource plans prior to 2015 included either a preferred plan or a recommended path forward. The 2016 integrated resource plan does not have a preferred plan or a recommended path forward, as Dominion did not have enough time to analyze a future in which either the Clean Power Plan implementation is delayed or a different form of carbon dioxide regulation is promulgated. Instead, Dominion intends to study these five plans that represent plausible future paths for meeting electricity needs while responding to changing regulatory requirements.
- All of the studied plans include:
  - 400 MW of utility-scale solar phased in from 2016-2020
  - 600 MW of solar generation from non-utility generators by 2017
  - 7 MW of solar from its "Solar Partnership Program"
  - 12 MW from the Virginia Offshore Wind Technology Advancement Project (VOWTAP) as early as 2018
  - Demand-side resources of 304 MW by 2031
  - 20-year extensions of four nuclear reactors by 2038
  - 1,585 MW of additional natural gas combined cycle capacity by 2019
- To show how the various plans can diverge from the previous year's plan, if Dominion were to adopt the most solar-focused plan, this plan projects 7,000 MW of additional solar resources by 2029, an increase of 3,500 MW over the solar-focused 2015 plan.[143]

To meet Colorado's multi-pollutant planning requirement, Xcel Energy submitted a plan that was approved by the Colorado PUC on December 9, 2010. Implementation of the plan will reduce NO_x levels 86 percent and CO_2 levels 28 percent relative to 2008 levels by 2018.[144] Black Hills Energy has also filed its electric resource plan (ERP). This plan includes the retirement of a coal-fired power plant and two older natural gas-fired gas units, as well as a proposal to build a 40 MW natural gas turbine. It plans to add 100 MW of capacity by 2017, and use competitive

---

[142] Ameren Missouri, *2014 Integrated Resource Plan* (Ameren Missouri, 2014). Available at: https://www.ameren.com/missouri/environment/renewables/ameren-missouri-irp.

[143] Dominion, *Dominion Virginia Power's and Dominion North Carolina Power's Report of its Integrated Resource Plan* (Dominion, April 2016). Available at: https://www.dom.com/corporate/what-we-do/electricity/generation/2016-integrated-resource-planning.

[144] Xcel Energy, *Xcel Energy-Emissions Reduction Plan*, (Xcel Energy, 2011). Available at: https://www.xcelenergy.com/staticfiles/xe/Corporate/Environment/10-12-303_CACJ-6E_FS.pdf.

bidding to meet the remaining 60 MW.[145] Work is well underway to implement Xcel Energy's emissions reduction plan under Colorado's Clean Air-Clean Jobs Act. Three coal units have been retired and new emissions controls finished in August 2014 have kept emissions rates below new permit levels.[146]

In Minnesota, projects currently implemented under the multi-pollutant legislation include the Minnesota Power's Arrowhead Regional Emissions Abatement (AREA) Project, Minnesota Power's Boswell 3 Emissions Reduction Plan, Xcel Energy's Mercury Reduction Plan, and Xcel Energy's Metropolitan Emissions Reduction Proposal (MERP). MERP, authorized in 2002, has shown an annual 93 percent reduction in SO_2, 91 percent reduction in NO_x, 81 percent reduction in mercury, 55 percent reduction in particulates, and 21 percent reduction in CO_2 from 2002 levels during the 2007 to 2009 time period.[147]

[145] Black Hills Energy, "Black Hills Energy Files Plan for Ongoing Reliable, Cost-effective Energy for Years to Come in Colorado." Available at: http://www.blackhillsenergy.com/node/34671#.UzHkuIXYhlt.

[146] Xcel Energy, *Colorado Clean Air-Clean Jobs Plan*. Available at: http://www.xcelenergy.com/Environment/Programs/Colorado_Clean_Air-Clean_Jobs_Plan.

[147] Xcel Energy, "Minnesota Metro Emissions Reduction Project – Environmental Benefits." Available at: http://www.xcelenergy.com/Environment/Programs/Minnesota_Metro_Emissions_Reduction_Project.

www.ingramcontent.com/pod-product-compliance
Lightning Source LLC
Chambersburg PA
CBHW081305170526
45165CB00011B/3419